FPGAs 101

FPGAs 101

Everything you need to know to get started

Gina R. Smith

ELSEVIER

AMSTERDAM · BOSTON · HEIDELBERG · LONDON
NEW YORK · OXFORD · PARIS · SAN DIEGO
SAN FRANCISCO · SINGAPORE · SYDNEY · TOKYO
Newnes is an imprint of Elsevier

Newnes

Newnes is an imprint of Elsevier
30 Corporate Drive, Suite 400, Burlington, MA 01803, USA
The Boulevard, Langford Lane, Kidlington, Oxford, OX5 1GB, UK

Notices

Knowledge and best practice in this field are constantly changing. As new research and experience broaden our
understanding, changes in research methods, professional practices, or medical treatment may become necessary.

Practitioners and researchers must always rely on their own experience and knowledge in evaluating and using
any information, methods, compounds, or experiments described herein. In using such information or methods
they should be mindful of their own safety and the safety of others, including parties for whom they have a
professional responsibility.

To the fullest extent of the law, neither the Publisher nor the authors, contributors, or editors, assume any liability
for any injury and/or damage to persons or property as a matter of products liability, negligence or otherwise, or
from any use or operation of any methods, products, instructions, or ideas contained in the material herein.

Library of Congress Cataloging-in-Publication Data
Smith, Gina R.
  FPGAs 101: Everything you need to know to get started / Gina R. Smith.
    p. cm.
  Includes bibliographical references and index.
  ISBN 978-1-85617-706-1 (alk. paper)
   1. Field programmable gate arrays. 2. Programmable array logic. 3. VHDL (Computer hardware
description language) 4. Digital electronics. I. Title.
  TK7895.G36S6525 2010
  621.39'5–dc22

                                                                            2009041496

British Library Cataloguing-in-Publication Data
A catalogue record for this book is available from the British Library.

ISBN: 978-1-85617-706-1

> For information on all Newnes publications,
> visit our website, www.elsevierdirect.com

10 11 12 9 8 7 6 5 4 3 2 1

Printed in the United States of America

Contents

About the Author

Gina R. Smith is the CEO and founder of Brown-Smith RDL Inc., located in Maryland. She is a Senior Electrical/Systems Engineer with an Associate's degree in Applied Science in Electronic Engineering Technology, a B.S. degree in Electrical Engineering, magna cum laude, and an M.S. degree in Systems Engineering with honors, from Johns Hopkins University. Through her company, Gina provides systems engineering and FPGA consulting services. Her accomplishments include a Technical Merit Award from Northrop Grumman, for one of her high-voltage designs. She wrote *The Art of FPGA Construction*, published in *Embedded Systems Design*, January 2008. She is a member of the International Council on Systems Engineering, Women in Defense, and National Defense Industrial Association. She has numerous years experience as a design and project engineer and technical leader. She worked in the fields of avionics, chemistry and biology, primary and secondary surveillance radar systems, information, friend or foe technology, and various other military and defense systems.

Gina has one daughter, Rebecca C. Smith, and lives in Maryland. She enjoys snow skiing, NASCAR, hiking, cooking, and fishing.

Gina can be reached via email at gina.smith@bsrdl.com.

Acknowledgments

I thank my daughter, Rebecca C. Smith, who is always so patient, loving, supportive, and understanding with me. She is my true source of motivation. I also thank my mother, Rebecca B. Smith, sisters Julie P. Webb and Sharon C. Smith, and my brother Maurice R. Smith for all their support and encouragement during this project. I am truly an engineer and not a writer. Without the support of my daughter and family, this book journey would have been almost impossible. Many thanks for the technical support I received from Mike Hines, Livia Castrucci, Christopher Loberg, Lawrence Wilson at Tektronix, Scott Silver at The Moving Pixel Company, and Brian Jacobsen at Synopsys for getting me a temporary license. Thanks to Xilinx, Altera, and ModelSim for letting me use their material.

I dedicate this book to my dad, who was my hero and number 1 supporter. He is one of "heaven's angels."[1] I love and miss him very much.

---

[1] James Wellington Smith, June 4, 1932–July 21, 2007

About This Book

This book describes the field programmable gate array (FPGA) development process in five development phases: design, synthesis, simulation, implementation, and programming. Each phase is presented in an easy-to-read and easy-to-understand format with examples, helpful tips, and step-by-step tutorials for the synthesis, implementation, simulation, and programming phases. The reader is provided Web addresses for the tools used in the tutorials. This book makes it easy for beginners to learn and understand how to create, modify, and work with FPGA designs. Experienced engineers will find it to be a good reference manual. A primer and some simple digital designs have been provided for those with no programming knowledge. It provides some basic information about writing, reading, and understanding high-level design languages, coding, and other tips. The primer may not be necessary for all readers, but as my mother always says, "It is better to have it and not need it than need it and not have it."

Acronyms

ABEL	advanced Boolean equation language
AHDL	Altera hardware description language
ALM	adaptive logic module
B.E.S.T	behavior extracting synthesis technology
BGA	ball grid array
CAN	controller area network
CIC	cascaded integrator comb
CLB	configurable logic block
CTRL	control
DFF	D flip-flop
DoD	Department of Defense
DRC	design rule check
DSP	digital signal processor
EDIF	electronic digital interchange format
FBGA	fine-pitch ball grid array
FFT	fast Fourier transform
FIR	finite impulse response
FPGA	field programmable gate array
FSM	finite state machine
GTL	gunning transceiver logic
GUI	graphical user interface
HDL	hardware description language
HSTL	high-speed transceiver logic
IEEE	Institute of Electrical and Electronics Engineers
IFF	information, friend or foe
I/O	input/output
IOB	I/O block
IOE	I/O element
IP	intellectual property
ISE	Integrated Software Environment

ISP	in-system programming
JTAG	Joint Test Advisory Group
LDT	lightning data transport
LED	light emitting diodes
LSB	least significant bit
LUT	look-up table
LVCMOS	low-voltage CMOS
LVDS	low-voltage differential signaling
LVTTL	low-voltage transistor-transistor logic
MSB	most significant bit
NCD	native circuit description
NGC	native generic compiler
NGD	native generic database
OTP	one-time programming
PAR	place and route
PCI	peripheral component interconnect
PROM	programmable read only memory
QoR	quality of results
RAM	random access memory
RTL	register transfer level
SPI	serial peripheral interface
SRAM	Static random access memory
Std	standard
TCK	test clock
TDI	test data in
TDO	test data out
TMS	test mode select
TQFP	thin quad flat pack
TRST	test reset
VHDL	very high speed integrated circuit hardware description language
XST	Xilinx Synthesis Technology

CHAPTER 1

Getting Started

1.1. Introduction

This chapter is a primer that provides beginners with some background information that will help in understanding the field programmable gate array (FPGA) development process described in this book. The FPGA process can be confusing and frustrating, making it more difficult to learn or understand, especially if you do not have at least a basic understanding of some background concepts. So, it is my hope this primer will fill in some, if not all, of the gaps. I am a firm believer that we can always learn something new; so for experienced engineers, I believe this book will be both a good refresher and an opportunity to learn something new. The chapter also provides some helpful hints and tips that I found to be useful over the years. I hope they will prove beneficial to you.

Some basic examples are provided to help you better understand hardware description language (HDL) and the FPGA development process. This primer is not meant to teach you all the ins and outs of writing HDL code but to help you better understand some of the terminology as you read the later chapters on the FPGA development process.

In this chapter, you will learn

- HDL coding.

- Tips for writing code.

- HDL editor features.

- HDL file structure.

1.1.1. VHDL

VHDL is a high-level hardware description language used to describe digital circuits that can be programmed into an FPGA. It is a softwarelike programming language that some people, myself included, refer to as *firmware*. It was developed based on a need by the U.S. Department of Defense (DoD). In 1987, the Institute of Electrical and Electronics

Doi:10.1016/B978-1-85617-706-1.00001-1

Engineers (IEEE) adopted VHDL as a standard, which was released as IEEE Standard (Std) 1076–1987 or VHDL-87. About every five years, the IEEE Standards Committee is supposed to reconvene to review, enhance, and make other modifications to the language. VHDL is also available as VHDL-93, VHDL-2000, VHDL-2002, and VHDL-2008.

VHDL is not case sensitive and requires no special formatting, such as spaces, tabs, or indentations. Each line of code or statement must end with a semicolon, ; . Filename extensions can be either .vhd or .vhdl.

1.2. Reserved Words

Reserved words are words that are defined by the language. Of the many reserved words, you will use some more often than others. Some common VHDL reserved words are shown in Table 1–1. Because VHDL is not case sensitive, reserved words may be in any case.

Table 1–1: VHDL Reserved Words

Reserved word	Description
All	References what precedes the .all
And	Logic AND function
Architecture	Secondary design unit
Begin	Signifies the start of sequential statements
Bus	A signal mode that has multiple drivers or signal bits
Case	Creates a multiplexer for a signal
Component	Used to define a component
Constant	Fixed signal value
Downto	Defines range of values
Else	Precedes alternate action following the "If-Then" statement
End	Signifies the end for many things, like entity, architecture, and If-Then statements
Entity	Primary design unit
If	Precedes initial conditional
In	Input signal port
Inout	Bidirection signal port
Is	A connective in a variety of statements
Map	Maps or connectors actual signal parameters
Not	Logic NOT function
Then	Used for conditional statements
Type	Enumerated type allows user to define data values
Or	Logic OR function
Others	Shortcut used to define all values in a range
Out	Output signal port
Port	Used for interface definition

(Continues)

Table 1–1: Cont'd

Reserved word	Description
Process	Group of sequentially executed code
Read	Allows an external file to be read
Std_logic	Signal type defining a single bit
Std_logic_vector	Signal type defining multiple bits
Signal	Used to assign an object a signal name and data type
To	Used in the middle of some keywords like range and downto
Write	Allows you to write to an external file

1.3. Tips for Writing Good Code

Over my many years as a digital designer, I had the opportunity to write many lines of code as well as review, modify, and inherit others' code. Through these sometimes hard experiences, I have developed some tips for writing good code. Some of the tips come from trying to remember code I previously wrote or understanding someone's code. As you become more experienced, you will discover tips that make writing, modifying, and reviewing code much easier.

1.3.1. Tip 1. Use Comments to Convey Information about the Code

Comments are a very important part of coding. You should provide comments as a way of conveying pertinent information about the line or section of code, see Example 1–1.

■ Example 1–1. Good and Bad Code Comments

```
Count <= NumberOfBaskets;        -- number of baskets = 5
                                 -- count equals number of baskets
                                 -- count <= NumberOfBaskets;
```

Good Comment.

Number of baskets = 5 is a good comment, because it defines the actual value being assigned to Count. There is no need to search through the code or another file to find the actual value.

Bad Comment.

In Count equals number of baskets, the comment states the obvious and provides no additional information. You have no way of knowing the actual value without further research.

You may find it hard to believe, but I have actually reviewed code that had comments like count <= NumberOfBaskets. The comment was the actual code commented out. I guess the coder did not really know what commenting meant.

■

It is so easy to remember all the ins and outs of your code while you are developing it but not so easy if you have to revisit the same code several weeks or months later. It can be even more difficult when you try to understand someone else's code.

Some would say that each line of code should have a comment. My personal opinion is that obvious code needs no comments. However, you should never generate code without some comments.

VHDL comments are preceded by double dashes: – –. The double dashes denote the start of a comment and continue until a carriage return is encountered.

There will be times when you will thank yourself or someone else for providing good comments.

1.3.2. Tip 2. Indent for Clarity and Readability

Indent using spaces or tabs to align groupings of codes. This makes it much easier to read the code and identify common groups of code. See Example 1–2.

1.3.3. Tip 3. Use Standard Format Convention

Standard format convention means that reserved words and user-defined names are presented in the same format. Some companies predefine coding conventions for writing code. If this is not the case, you may decide that all reserved words will be in upper case and user-defined names in lower case. This makes it easy to immediately identify reserved words from user-defined signals. See Example 1–3 for some suggested format conventions.

1.3.4. Tip 4. Include a Header Section

The header section is an optional section that you should include prior to your code. This section may contain information such as the author's name, date created, filename, a brief description summarizing the design, and revision history. You are free to include whatever information you feel will be beneficial. Example 1–4 shows one possible outline for the header section.

■ Example 1–2. Indention

Not Indented	Indented
If count = '100' **Then**	**If** count = '100' **Then**
count = '1' ;	count = '1' ;
Else	**Else**
count <= count + '1' ;	count <= count + '1' ;
End If;	**End If;**

■ Example 1–3. Suggested Format Conventions

Capitalize the first letter; for example, `Signal`.
Lower case all letters; for example, `signal`.
Upper case all letters; for example, `SIGNAL`.
Upper and lower case to separate words; for example, `StartCounter`.
Underscore to separate words; for example, `Start_Counter`.

■

■ Example 1–4. Optional VHDL Header Section

```
--************************ Header Section ************************
-- Name          :
-- Date          :
-- Filename       :
-- Description    :
-- Revision History
-- Date              Initials           Description
--*********************** End Header Section **********************
```

■

1.3.5. Tip 5. Use Brief Descriptive Names

Always use *brief* but *descriptive* names. Descriptive names make the code easier to read and understand. A good descriptive name should provide information about a signal's function, see Example 1–5.

■ Example 1–5. Descriptive User-Defined Names

You need to name a 40 MHz clock signal.

Descriptive signal name: `clock40Mhz` or `clk40MHz`.

At first glance, anyone reading the code will know this signal is a 40 MHz clock.

Nondescriptive signal name: `c40Mhz` or `c40`.

At first, second, and third glance, it would be very difficult to know anything about the signal based on the nondescriptive signal name, unless there were a comment defining the signal. While comments are very important, they should not be used as a substitute for good signal names.

■

A lot can be said about being brief with your signal names. One good reason for not having long signal names is that, the more you type the name, the more you increase the chance of typos. There is no hard and fast rule as to what is too long, but keep in mind you can use abbreviations to shorten a name.

Once I inherited a coworker's code and *all* the signal names were only a single letter (i.e., a, b, c, etc.) with no comments to be found in the entire design. It took me a while but I finally got the code straightened out and that is how this became one of my "tips." For that code, the signal names were too brief and not descriptive.

1.4. HDL Text Editors

Having a good editor is really important because it is the tool you use to develop and edit code. It should be something that you find easy to use, and it should have HDL features, such as language templates or syntax color highlighting. Many of the FPGA development tools offered by manufacturers like Xilinx and Altera include a text editor. Standalone editors are available for free or purchase. I found some really good free editors just by searching the Internet. My personal preference is a standalone editor. There are many standalone editors. My advice is to make sure that the standalone editor provides support for HDL.

1.4.1. Standalone Text Editor

HDL Works offers Scriptum, a free text editor that supports VHDL and Verilog on Windows and UNIX platforms. I found this editor to be easy to use, with a lot of helpful features. Some of its features are

- Language templates.

- Syntax coloring.

- Multiline comment and uncomment.

- Column and row select/edit.

- Change of case for selected reserved words.

- Bookmarks.

- Standard search, find, and replace.

HDL Works Scriptum text editor can be downloaded for free at www.translogiccorp.com/index.html.

1.4.2. Fee-Based Text Editor

Symphony EDA offers both free and fee-based text editors. The editors are a part of its simulation/debug software package. The fee-based edition offers more features than the free editor. It has some of the same standard features, which include

- Language templates.

- Syntax coloring.

- Multiline comment and uncomment.

- Column and row select/edit.

Symphony EDA can be found at www.symphonyeda.com.

Downloading and evaluating different editors is an excellent way to try the editor before you buy it. Who knows, you may find a free editor is sufficient for your needs. I did.

Although not required, I highly suggest selecting an editor that is on the same platform (Windows or UNIX) as your FPGA development tools. As simple as this sounds, when I first started with VHDL and FPGA development, some of the tools were Windows based and others were UNIX. We had to ftp the files between the two systems. It was so confusing and created such a mess. When the lead engineer left the company, I jumped at the chance to have all the applications on one platform.

1.5. Editor Features

VHDL code is just a text file, meaning you can use any text editor to create your design. However, it is best to use a text editor that provides special HDL coding features, such as syntax color highlighting, language templates, row/column editor, comment/uncomment selected text, indent/unindent selected text, and predefined keyword font convention. There are many different editors, offering various features, so evaluate a few to determine the best fit for you. Following are some of the features often offered and beneficial during developing and editing code.

1.5.1. Syntax Color Highlighting

This occurs when syntax items are displayed or highlighted in a specific color. Syntax items can be keywords (sometimes defined as various levels or categories), regular text, comments, variables, or strings. They can vary from editor to editor. Generally, the syntax color highlight is set to a default value; however, many editors allow the user to redefine the colors. See Example 1–6 for a snapshot color highlight, where bold and italics represent specific colors.

■ Example 1–6. Syntax Color Highlighting

```
If count = '100' Then
   count <= count + '1';          -- this is a comment
End If ;
```
Notice the keywords **If** , **Then**, and **End If** are bold; while the comment -- *this is a comment* is italicized.

■

Syntax highlighting makes it easy to quickly identify specific code elements such as keywords and comments. While some editors offer more syntax highlight items than others, I consider it a must-have feature.

1.5.2. Language Templates

The language template presents HDL syntax for specific language functions in a fill-in-the-blank format. For beginners, code templates can be a lifesaver. For example, if you need to know how to write an "If-Then-Else" statement but cannot remember or do not know the syntax, then you may be provided a fill-in-the-blank template similar to the one shown in Example 1–7.

■ Example 1–7. Language Template: If-Then-Else

```
If <insert condition> Then
     <insert action(s) >;
Else
     <insert alternation action(s)>;
     End If;
```

■

Now all you have to do is insert your code where indicated by the placeholders on the template. Templates vary from editor to editor, but they have the same basic concept.

1.5.3. Row and Column Editor

Most people are familiar with row editing; however, column editing is not as familiar. A row editor is used when multiple rows are selected at one time, see Example 1–8. Similarly, a column editor allows you to select multiple columns on different rows at once, see Example 1–9. Column editing is a great feature to use when the data you want to edit is in the same column but on different rows. For most HDL editors, the alternate,

■ Example 1–8. Row Edit

Select, copy, and paste the second row. The gray area indicates the text selected using the row editor.

Row edit copy:
```
If front_door_open = '1' Then          -- 1 means front door was opened
   alarm_timer <= alarm_timer + '1' ;   -- increment time to sound alarm
End If ;
```

Row edit paste:
```
   alarm_timer <= alarm_timer + '1' ;   -- increment time to sound alarm
```
Notice that only the selected row was copied and pasted.   ■

■ Example 1–9. Column Edit

Use the column edit to select, copy, and paste the two comments. The gray area shows the selected text.
```
If front_door_open = '1' Then          -- 1 means front door was opened
   alarm_timer <= alarm_timer + '1' ;   -- increment time to sound alarm
End If ;
   Column edit paste:
          -- 1 means front door was opened
          -- increment alarm counter
```
With column editing, I am able to copy and delete selected data without affecting the surrounding text, but I have little success with pasting.   ■

alt, key with the mouse button is used for column editing; however, in Microsoft Word, it is the control, Ctrl, key with the mouse.

1.5.4. Comment/Uncomment Selected Text

Sometimes it is necessary to comment out multiple lines of code instead of individually commenting each line. Some editors provide an option to comment/uncomment select lines. Some editors comment/uncomment only at the beginning of a row. This means that, if the cursor is put on any part of a row, the comments syntax or double dashes (– –) for VHDL, see Example 1–10, are inserted as the first two characters on that row. When uncommenting, some editors delete only the comment syntax at the beginning of the row, ignoring any other comment syntax on the row, see Example 1–11.

■ Example 1–10. Comment

Select and comment to the three lines of code. The gray area shows selected text.

```
If front_door_open = '1' Then          -- 1 means front door was opened
    alarm_timer <= alarm_timer + '1';   -- increment time to sound alarm
End If;
```

The selected text is now commented.

```
-- If front_door_open = '1' Then         -- 1 means front door was opened
-- alarm_timer <= alarm_timer + '1';      -- increment time to sound alarm
-- End If;
```

■

■ Example 1–11. Uncomment Text

Select text to be uncommented. The gray area shows the selected text.

```
-- If front_door_open = '1' Then         -- 1 means front door was opened
-- alarm_timer <= alarm_timer + '1';      -- increment time to sound alarm
-- End If;
```

Selected text is now uncommented.

```
If front_door_open = '1' Then          -- 1 means front door was opened
    alarm_timer <= alarm_timer + '1'; -- increment time to sound alarm
End If;
```

■

■ Example 1–12. Indent Text

Select and indent the second and third lines. The gray area shows the selected text.

```
If front_door_open = '1' Then          -- 1 means front door was opened
    alarm_timer <= alarm_timer + '1';   -- increment time to sound alarm
End If;
```

The second and third lines are indented.

```
If front_door_open = '1' Then          -- 1 means front door was opened
    alarm_timer <= alarm_timer + '1';   -- increment time to sound alarm
    End If;
```

■

■ Example 1–13. Unindent Text

The **End If;** should be aligned under If. Now select and unindent the last line. The gray area shows selected text.

```
If front_door_open = '1' Then          -- 1 means front door was opened
   alarm_timer <= alarm_timer + '1' ;   -- increment time to sound alarm
   End If;
```

The third line is now properly aligned.

```
If front_door_open = '1' Then          -- 1 means front door was opened
   alarm_timer <= alarm_timer + '1' ;   -- increment time to sound alarm
End If;
```

■

1.5.5. Indent/Unindent Selected Text

Some editors allow you to indent only a portion of the row, while others indent the entire row. This feature generally works like the comment/uncomment feature. Example 1–12 shows selected text being indented, and Example 1–13 shows how selected text is unindented.

1.5.6. Predefined Font Convention

The predefined font convention is when the editor converts keywords or selected text to a specific font style or size, such as all lower or upper case. This feature can be used to keep your code consistent, because sometimes you may miss applying your font convention during the development process. Some editors may require you to highlight the text you want to convert, while others perform it automatically. In Example 1–14, the line(s) or code are highlighted for the font conversion.

■ Example 1–14. Font Convention

Convert keywords to upper case.

```
if front_door_open = '1' then          -- 1 means front door was opened
   deactivate_alarm <= deactivate_alarm + '1' ; -- increment alarm counter
end if;
```

Upper case keywords converted.

```
IF front_door_open = '1' THEN          -- 1 means front door was opened
   deactivate_alarm <= deactivate_alarm + '1' ; -- increment alarm counter
END IF;
```

■

1.6. Signals

In VHDL, signals represent some kind of data. They are assigned a name and data type. The basic signal syntax follows:

Signal <signal name>: <data type>;

More advanced signal assignments are possible but not discussed in this book.

You can use signals in mathematical equations, to assign values, to connect other signals, and to store values in them. They must be assigned unique, nonreserved word names and a data type.

1.6.1. Signal Data Types

The VHDL data are of a specific type such as std_logic, std_logic_vector, bit, bit_vector, or user defined. Std_logic is read as standard logic and std_logic_vector as standard logic vector. Bit and bit_vector are read as written. The user-defined type is when the coder defines the signal type. This is a little more advanced and can be somewhat confusing when you are first starting out, so it is not covered in this book. Once you are more comfortable with the language, it will be easier to understand and implement the more advanced aspects of the language. Plus std_logic and std_logic_vector are generally the most commonly used data types. Each signal type has acceptable values. There are nine acceptable values for std_logic and std_logic_vector, see Table 1–2.

Std_logic signals represent one data bit and std_logic_vector represents several data bits. The signal assignments for standard logic and standard logic vector data types are shown in Example 1–15. The number of data bits for a std_logic_vector is defined in the signal assignment statement.

Table 1–2: Standard Logic Acceptable Values

Value	Description
0	Low or logic zero
1	High or logic one
W	Weak unknown signal
L	Weak low
H	Weak high
U	Unknown or uninitialized
Z	High impedance
X	Unknown
-	Don't care

■ Example 1–15. Standard Logic and Standard Logic Vector Signal Assignment

```
Signal      clock_in         : std_logic;
Signal      up_counter       : std_logic_vector (4 downto 0);
```

■ Example 1–16. Valid Standard Logic Signal Values

```
clock_in value is "1"
up_counter value is "1XZ0U".
```

The signal named `clock_in` has a data type of `std_logic`. This means `clock_in` can have only one of the nine acceptable values, while the signal named `up_counter` is 5 data bits wide and each of the bits can be one of the nine acceptable values. The most significant bit (MSB) is bit 4 and the least significant bit (LSB) is 0. See Example 1–16 for valid signal values.

`Bit` and `bit_vector` have two acceptable types, see Table 1–3.

`Bit` represents one data bit, and `bit_vector` represents several data bits. Example 1–17 shows signals `clock_out` being assigned data type bit and `down_counter2` bit vector.

The signal named `clock_out` has a data type of bit. This means `clock_out` can have only one of the two acceptable values, while the signal named `down_counter2` is 4 bits wide and each of the bits can be one of the two acceptable values. See Example 1–18 for valid bit signal values. The MSB is bit 3 and the LSB is 0.

It is okay to use `bit` and `bit_vector`; however; they are rarely the deserved data type.

Table 1–3: Bit Acceptable Values

Value	Description
0	Low or logic zero
1	High or logic one

■ Example 1–17. Bit Signal Assignment

```
Signal      clock_out        : bit;
Signal      down_counter2    : bit_vector (3 downto 0);
```

■ **Example 1–18. Valid Bit Values**

```
clock_out is "0"
down_counter2 is "0011"
```

■

Table 1–4: VHDL Signal Name Restrictions

Rule	Acceptable	Not Acceptable
Must start with letter	four	4
Cannot be a keyword	input_signal	input
Don't use special characters	input_data	$id
Must not contain spaces	InputData	Input Data

1.6.2. Signal Names

Signal names are user defined, but VHDL has some name restrictions, such as those provided in Table 1–4. In addition to the VHDL restrictions, remember to make your names descriptive.

1.7. File Structure

The VHDL file structure consists of three sections: the library declaration, entity section, and architecture section with an optional header section.

1.7.1. Optional Header Section

As stated before, the header section is completely optional but highly suggested. However, I found the revision history to be beneficial, especially when modifying someone else's code. When I fix code problems, the revision history enables me to see if the current problem existed before any previous code changes. I have mixed feeling about keeping the prereleased code revision history in the released code. It is a good idea to have the revision history while developing, just in case someone else inherits the code. However, for the released code, this history may not add any value, in which case should be removed. It all depends on the specific situation.

1.7.2. Library Declaration

Just like software code, VHDL must be compiled. The place where the compiler stores the design information and other files to be used for analysis, synthesis, and simulation is called the *library*. The library declaration section is where you declare or call out libraries. By using the library clause you make the library visible and its contents available to the design. The use clause, which follows the library clause, states which package from the library to use. Libraries can be a defined by standards, users, or third parties, like manufacturers.

IEEE is a commonly used standard library. Some of the packages in IEEE are `Std_logic_1164` and `Std_logic_arith`, see Table 1–5 for some of the data types and functions defined by these packages.

User-defined libraries are those created by regular users or designers. Oftentimes design groups or projects utilize user-defined libraries by storing common constants, data types, and other commonly used things in a library shared by the group. This can save a lot of time, because each individual is not spending time creating the same information. The user-defined library is placed in a common, team-accessible area. Another benefit to groups using a user-defined library is that it ensures everyone uses the same values, functions, definitions, and the like. This does not guarantee the values are correct; however, it makes it easier to correct something wrong in one place rather than in several files.

Third party libraries are supplied by companies like Xilinx and Altera. These libraries contain such information as timing used for simulation, IP cores and logic gates.

The library syntax is the reserved word *Library* followed by the library's name. The "use" clause syntax specifies the package, its library, and how much of the package is used, see Example 1–19. A package is a separate VHDL file that defines things like functions, data types, constants, and procedures.

Table 1–5: IEEE Standard Library Packages, Data Types, and Functions

Package	Data Types	Functions
Std_logic_1164	std_logic, std_logic_vector, std_ulogic, std_ulogic_vector	AND, NAND, OR, NOR, XNOR, NOT
Std_logic_arith	Unsigned, signed	+, -, *, ABS, <, <=, =, >=, >

■ Example 1–19. Library and Use Syntax

Library <library name>;
Use <library name>.<package>.<what portion are you using>;
   For example, `std_logic` is a widely used data type, which is defined in the
   `std_logic_1164` package. To use this data type, you need to declare the library
   where it is defined and state the package using the use clause in the library declaration

(Continues)

section, see Example 1–20. Some additional packages included in the IEEE library are `std_logic_arith`, `std_logic_unsigned`, and `std_logic_signed`. If you are using several packages from the same library, the library needs to be stated only once.

What these two lines say is this: Make the IEEE library visible to the design and make the entire (i.e., `all`) `std_logic_1164` package available to your design. So whatever is defined in the `std_logic_1164` package can now be used in your code. You are not required to use all the features in the package and can specify only the portion you want: however, using the `all` just makes things easier. So my advice is, unless you have a good reason for not wanting to include everything, it is a good idea to stick with `all`. Now you need not worry about changing the use statement if your design requires additional features.

■

1.7.3. Entity Section

The entity section is where you define all the inputs and outputs. The syntax for the entity section is shown in Example 1–21, and Example 1–22 shows a simple design entity.

Note: Entities can be a little more complex; however, for this primer, the entity is the top level of the design and represents the inputs and output pins on the FPGA.

■ Example 1–20. Library and Use

```
Library IEEE;              -- IEEE library is visible to the design
Use IEEE.std_logic_1164.all;  -- The contents in the std_logic
                           -- package can now be used
                           -- in the design code
```
■

■ Example 1–21. Entity Syntax

```
Entity <entity name> Is Port(
   <signal name   : <signal direction> <data type>);
End <entity name>;
```
■

■ Example 1–22. Entity Code

```
Entity test_code Is Port(
  clk           : In std_logic;    -- input clock
  start_counter : In std_logic;    -- starts counter when door is opened
  ready_signal  : Inout std_logic; -- indicates alarm status
  sound_alarm   : Out std_logic);  -- alarms buzzer when timer expires
End test_code;
```

■

1.7.4. Architecture Section

The architecture section is where you write the design code, see Example 1–23. The design code describes the functions by using the software-like programming language VHDL.

Now, you use the defined functions, data types, and so for the package(s) declared or called out in the library declaration section. Your design receives and passes design data using the input, bidirectional (inout), and output ports defined in the entity section.

■ Example 1–23. Architecture Syntax

```
Architecture <architecture name> Of <entity name> Is
<Define signals and constants>
Begin
```

This section is where the design is written. It consists mainly of component instantiations, synchronous logic, sequential statements, processes, concurrent statements, and asynchronous logic.

Component instantiation basically makes a direct connection to a library component. The actual code for the component is predefined in another file. More details about component instantiation are provided in the testbench section.

Synchronous logic is code that gets updated based on an event, such as the rising or falling edge of a clock.

Sequential statements are found in processes and executed in the order in which they appear.

A *process* is a group of code that is executed sequentially. They are like mini programs with very specific format, which includes the use of the reserved words `process`, `begin`, and `end process`, and a `sensitivity list`, see Example 1-24.

■

■ Example 1–24. Process Syntax

```
<process name>: Process (sensitivity list)
Begin
        <sequential statements>
End Process ;
```
■

It is good coding practice to perform only one function in a process. The code inside a process is executed only when any signal in the sensitivity list changes state.

The process name is optional and user defined. Even though it is optional, you should always name your processes. The name should be short and descriptive enough to allow you to distinguish one process from another.

One good reason to make the name descriptive is that you may have a design with four counters, so naming them counter1, counter2, . . ., tells you nothing about the counters. Let us say one counter counts the number of times the temperature sensor goes below 32°F, above 95°F, stays at 0°F for longer than 5 minutes and another one at 60°F for longer than 10 minutes. You may decide to name them count_below32F, count_above95F, temp_at0F, and temp_at60F.

The sensitivity list is where you list all the signals that you want to cause the code in the process to be evaluated whenever it changes state. For example, clock or master reset is often used in a sensitivity list. Whenever the reset or clock changes state, the code inside the process is executed.

Concurrent statements are outside of processes and executed or updated at any time any of the signals changes, see Example 1–25.

■ Example 1–25. Concurrent Statement

```
Sum_Temp       <= count_above95F + count_below32F;
Sum_Temp changes and is updated any time count_above95F or count_below32F
   changes.
```

   Asynchronous logic is updated or changed independent of events.

End <architecture name>;
■

Now that you have all the pieces, Example 1–26 shows how it looks all put together. Using this or a similar template is a good way to start each design.

▪ Example 1–26. VHDL File Structure

```
--************************* Header Section *************************
-- Name            :
-- Date            :
-- Filename        :
-- Description     :
-- Revision History
-- Date              Initials            Description
--
--*********************** End Header Section ***********************
Library IEEE;
Use IEEE.std_logic_1164.all;

Entity <entity name> Is Port
   (<list of ports or design inputs and outputs>);
End <entity name>;

Architecture <architecture name> Of <entity name> Is
   <in this section define signals and constants>

Signal <signal name>           : Data Type;

Begin
<concurrent statements>

<process name>: Process (sensitivity list)
Begin
   <sequential statements>
End;

End <architecture name>;
```

▪

So far the entity defines only the design's interface; however, your design most likely requires additional signals. These signals are defined in this section, prior to them being used and the actual design code. Once this is complete, you can take the signal names and the available features from the packages and develop code using asynchronous, synchronous, concurrent, and sequential code.

1.8. Starter Tips

A lot is involved in developing FPGA designs. Here are some tips to help as you get started.

- Utilize field application engineers and salespeople, who can really provide some good help and guidance. They can provide information on the latest hardware and tool developments, suggest devices, clarify details about their products and services, provide samples, arrange demos and software/hardware trials and temporary licenses, and many other things. If they cannot help you, they will at least point you in the correct direction. I realize some salespeople can be pushy, but I have found most of them to be very helpful and not very pushy.

- Evaluation boards are a good way to get experience. The evaluation boards are offered by manufacturers and third party companies. They come with different FPGA devices, development software, programming and other cables, power supplies and other resources like light emitting diodes, LEDs, pushbuttons, switches, and oscillators. Check out the specific manufacturer's Web site for specific details on the board and suppliers. Some companies purchase evaluation boards to try specific devices or features before using them in designs. So, if you are considering using an FPGA that has an embedded processor, then it may be a good idea to purchase an evaluation board and try it before using it in a design.

- Take advantage of the many free and trial offers. This will give you a good opportunity to experiment and discover your likes and dislikes.

1.9. Chapter Overview

Everyone has his or her style and approach to FPGA development. As you work with it, you will develop your style. So do not worry if you see things differently from everyone else; the process should be the same. Once you learn the general process you will have no problem using your knowledge to develop and understand designs and switching between different development tools and manufacturers.

Key Points to Remember

- Select an editor that you find easy to use and that has features you like. You will be using your editor a lot to develop and modify code.

- Take the time to include the optional headers section prior to your design code. Use this section to briefly describe the design and provide revision information.

- Always include "good," meaningful comments in your code.

- Make your user names brief and descriptive.

Chapter Links

For your convenience, here is a list of links to the editors discussed in this chapter:

HDL Works Scriptum located at www.translogiccorp.com/index.html.

Symphony EDA located at www.symphonyeda.com.

CHAPTER 2

Simple Designs

2.1. Introduction

The purpose of this chapter is to present some simple VHDL design code. These designs can easily be modified to perform advanced functions or be copied and used as a standalone design or a part of a larger design. In this section, the design code consists of a combination of processes, concurrent and sequential statements, and synchronous and asynchronous logic. When appropriate, the designs have a block diagram, corresponding to the VHDL design code and some brief comments to help explain the code.

In this chapter, you will learn

*   How to create simple VHDL designs.

*   How to add more complexity to simple designs.

*   VHDL shortcuts.

2.2. Starter Template

For each new design, I like to start with a "starter" HDL template. The starter HDL template consists of a header section, library declaration section, entity, and architecture syntax placeholders. Because I always use the IEEE library and the `std_logic_1164` package, I have made line 10 `Library IEEE;` and line 11 `Use.IEEE.std_logic_1164.all;` statements a part of my template instead of syntax placeholders. While many editors provide a starter HDL template, I present my template as an additional option.

When I first started writing VHDL, I could not remember if the signal name assignments went before or after the `Begin` in the architecture section. Therefore, a signal assignment was included as a placeholder in my template. The template can save you time, because it keeps you from retyping the same information from program to program. As you write code, you may decide to make your own template or modify a preexisting template. My starter HDL template is shown in Listing 2–1. Because the header section has no importance for the examples in this section, it is shown only in the first example.

Doi:10.1016/B978-1-85617-706-1.00002-3

Listing 2–1. VDHL Starter Template

```
1.     --************************Header Section ****************************
2.     -- Name              : Rebecca B. Smith
3.     -- Date              : August 25, 2009
4.     -- Filename          : Entity Name.vhd
5.     -- Description       : This starter HDL template provides placeholders and
   syntax that can be used
6.     -- : to help develop design code. Modify the template to meet your needs.
7.     Revision History
8.     -- Date                 Initials                Description
9.     --********************End Header Section ********************
10.    Library IEEE;                -- define library and packages needed for
   this design
11.    Use IEEE.std_logic_1164.All;
12.
13.    Entity <entity name> Is Port(        -- define interface signals
14.       <signal name>          : <direction> <data type>;
15.       <signal name>          : <direction> <data type>);
16.    End <entity name>;
17.
18.    Architecture <architecture name> Of <entity name> Is
19.    Signal <signal name>        : data type;        -- define internal signals
   if necessary
20.
21.    Begin
22.    <concurrent statements>          -- add concurrent statements if necessary
23.                                     -- statement(s) will update anytime
24.       <process name>: Process (sensitivity list)     -- add process if
   necessary
25.    Begin
26.       <sequential statements>;       -- statement updates when a signal
                   in sensitivity list changes
27.    End Process;
28.    End <architecture name>
```

2.3. Mathematical Functions

Mathematical functions, such as adder, subtractor, multiplier, and divider, are performed using arithmetic operators; see Table 2–1. The arithmetic operators can be used in concurrent and/or sequential statements; it depends on the circuit. To keep things simple, this example uses only a concurrent statement, and a sequential statement is demonstrated in later examples.

Table 2–1: Mathematical functions

Symbol	Data Types		Package
	Accepts	Returns	
+ Addition	std_logic, integer, std_logic_vector	std_logic_vector	std_logic_unsigned
	signed, unsigned, natural, integer	signed, unsigned	numeric_std
– Subtraction	std_logic, integer, std_logic_vector,	std_logic_vector	std_logic_unsigned
	signed, unsigned, natural, integer	signed, unsigned	numeric_std
* Multiplication	std_logic_vector	std_logic_vector	std_logic_unsigned
	signed, unsigned, natural, integer	signed, unsigned	numeric_std
/ Division	signed, unsigned, natural, integer	signed, unsigned	numeric_std

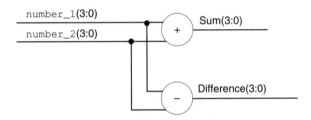

Figure 2–1: Adder and Subtractor

Figure 2–1 shows a simple two-input adder and subtractor. The design code is shown in Listing 2–2. If you are writing design code based on a schematic, it is best that the code and schematic signal names match. This makes writing and debugging the code much easier.

Lines 10–12. Library Declaration

The addition and subtraction operators are defined in the std_logic_unsigned package, which is made visible to the design in the library declaration section. You may not know all the libraries required at the beginning of your design, but you can add them as needed. Knowing which library to use can be challenging, until you remember which library contains the function(s) you need. So, at first, you may have to open different libraries to find what you need.

Listing 2–2. Adder and Subtracter

```
1.   --************************* Header Section *************************
2.   -- Name                :
3.   -- Date                :
4.   -- Filename            :
5.   -- Description         :
6.   -- Revision History
7.   -- Date                    Initials              Description
8.   --
9.   --*********************** End Header Section ***********************
10.  Library IEEE;
11.  Use IEEE.std_logic_1164.All;
12.  Use IEEE.std_logic_unsigned.All;
13.
14.  Entity MathematicalOperators Is Port (
15.      number_1         : In std_logic_vector (3 Downto 0);
16.      number_2         : In std_logic_vector (3 Downto 0);
17.      sum              : Out std_logic_vector (3 Downto 0);
18.      difference       : Out std_logic_vector (3 Downto 0));
19.  End MathematicalOperators;
20.
21.  Architecture arch_MathematicalOperators Of MathematicalOperators Is
22.
23.  Begin
24.  -- the sum and difference are calculated and provided as output for two
     4-bit numbers
25.      sum              <= number_1 + number_2;
26.      difference       <= number_1 - number_2;
27.  End arch_MathematicalOperators;
```

Lines 14–19. Entity Section

The interface signals for this design are defined as follows.

Two inputs:

Signal Names: number_1 and number_2

Data Type: std_logic_vector

Size: 4 bits

Two outputs:

Signal Names: sum and difference

Data Type: `std_logic_vector`

Size: 4 bits

Line 14 defines the entity's name as `MathematicalOperators`. This name was selected because it provides some detail about the design code's function. You should develop and use a naming convention for your entity, architecture, and filename. I found the naming convention to be useful, especially when working with multiple files. Using a name that gives an indication of the design code's function makes it easy to quickly identify files without having to open them.

Developing a standard naming convention can be tricky, especially when using different development tools, as they may have different filename restrictions. I have encountered tools that did not allow spaces in filenames, while another required my entity's name to be the same as the filename. So my naming convention for the entity does not include spaces and my filename is always the same as my entity. While this naming convention was based on older tool versions and those restrictions may no longer apply, I maintain this naming convention.

The input and output signals are defined as `std_logc_vector`, meaning they have multiple bits. The reserved word `Downto` (as used in this example) or `Upto` defines the range of data bits. The number before the keyword `downto` represents the most significant bit (MSB) and the number after it represents the least significant bit (LSB). In other words, `number_1(3 downto 0)` means `number_1` has 4 bits, where the MSB is `number_1(3)` and LSB is `number_1(0)`.

Lines 21–27. Architecture Section

This section contains the code that describes the design's functions. The output signal `sum` is assigned the result from adding together `number_1` and `number_2`. While the output signals `sum` and `difference` are concurrent statements (update any time `number_1` or `number_2` changes), they could have been sequential (i.e., inside a process and updated only when a signal from the sensitivity list changes).

2.4. Logic Gate

Logic gate circuits, such as OR, NOR, AND, and NAND, are implemented using the logic operators. The logic gate circuits shown in Figure 2–2 are represented by the design code shown in Listing 2–3.

Lines 16–19. Logic Operations

The output signals are assigned the result from performing the logic operation on the right side of the statement. These signal assignments are concurrent but could have been sequential statements.

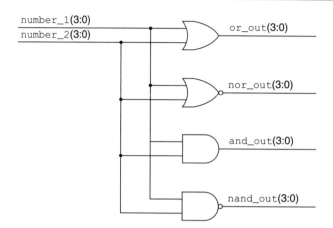

Figure 2–2: Logic Gates

Listing 2–3. Logic Gates

```
1.     Library IEEE;
2.     Use IEEE.std_logic_1164.all;
3.
4.     Entity LogicGates Is Port (
5.        number_1         : In std_logic_vector (3 Downto 0);
6.        number_2         : In std_logic_vector (3 Downto 0);
7.        or_out           : Out std_logic_vector (3 Downto 0);
8.        nor_out          : Out std_logic_vector (3 Downto 0);
9.        and_out          : Out std_logic_vector (3 Downto 0);
10.       nand_out         : Out std_logic_vector (3 Downto 0));
11.    End LogicGates;
12.
13.    Architecture arch_LogicGates Of LogicGates Is
14.    Begin
15.     -- this example illustrate how to implement logic gate code
16.        or_out           <= number_1 Or number_2;
17.        nor_out          <= number_1 Nor number_2;
18.        and_out          <= number_1 And number_2;
19.        nand_out         <= number_1 Nand number_2;
20.    End arch_LogicGates;
```

2.5. D Flip-Flop

A simple D flip-flop (DFF) is shown in Figure 2–3. The VHDL design code for the DFF is shown in Listing 2–4.

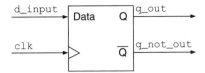

Figure 2–3: DFF

Listing 2–4. DFF

```
1.     Library IEEE;
2.     Use IEEE.std_logic_1164.All;
3.
4.     Entity Dff Is Port (
5.       reset          : In std_logic;
6.       clk            : In std_logic;
7.       d_input        : In std_logic;
8.       q_out          : Out std_logic;
9.       q_not_out      : Out std_logic);
10.    End Dff;
11.
12.    Architecture arch_Dff Of Dff Is
13.    Begin
14.
15.    q_not_out        <= Not (d_input);    -- inverted output of the DFF
16.
17.    dff_process: Process (reset, clk)
18.    Begin
19.      If reset = '1' Then
20.        q_out          <= '0';
21.      Elsif rising_edge (clk) Then
22.            q_out          <= d_input;    -- after the rising edge the output
   gets the value of the input
23.      End If;
24.    End Process;
25.    End arch_Dff;
```

Lines 17–24. DFF Process

Remember that the code inside a process is executed sequentially and only when a signal in the sensitivity list (i.e., `reset` and `clk`) changes state. The reset signal is asynchronous, so whenever it goes active or high, the outputs of the D flip-flop immediately (minus normal internal chip delays) goes low. Under normal operating conditions, when reset is inactive or low, on the rising edge of the clock, the input data is transferred to the outputs.

Line 22. DFF Output

The input data of the DFF is clocked or transferred to the output on the rising edge of the input clock, clk. This is a sequential operation, which is performed inside the process named dff_process, line 17.

Line 15. Inverted DFF Output

This is the inverted Q output from the DFF. The invert Q output only changes when Q changes on the rising edge of the clock.

Note: The reset signal shown on line 19 is used to set the outputs to a known or initial condition. This signal represents the power-on reset, it is not a part of the DFF, and does not appear on the symbol.

A synchronous enable can easily be added to this design by making some changes to the entity and the process in the architecture section. The entity change, shown in Listing 2–5, is the insertion of line 4. The process changes in the architecture section, shown in Listing 2–6, are lines 7 and 10.

Listing 2–5. Synchronous Enable DFF Entity Changes

```
1.      Entity DffSynEa Is Port (
2.        reset          : In std_logic;
3.        clk            : In std_logic;
4.        enable         : In std_logic;
5.        d_input        : In std_logic;
6.        q_out          : Out std_logic;
7.        q_not_out      : Out std_logic);
8.      End DffSynEa;
```

Listing 2–6. Synchronous Enable Process Changes

```
1.      DffSynEa_process: Process (clk, reset)
2.      Begin
3.        If reset = '1' Then
4.          q_out          <= '0';
5.          q_not_out      <= '1';
6.        Elsif rising_edge (clk) Then
7.          If enable = '1' Then           -- sync enable statement
8.            q_out      <= d_input;        -- after the rising edge the output
     gets the value of the input
9.            q_not_out  <= Not (d_input);   -- inverted output of the DFF
10.         End If;
11.       End If;
12.     End Process;
```

Line 4

Add input signal named `enable` to the entity.

Lines 7 and 10

Add an If -Then condition following rising edge statement in the `dff_process`; where `enable = 1` indicates active high or `enable = 0` indicates active low.

2.6. Latch

Sometimes you need to latch data. You can create latches by using the reserved word `When`. This example demonstrates only one of several uses for the reserved word `When`. The syntax for `When` is shown in Example 2–1.

The code for the latch shown in Figure 2–4 is shown in Listing 2–7.

■ Example 2–1. `When` **Syntax**

```
<output data signal name>        <= <input data signal name> When
                                 <latch condition>
                                 Else < output data signal name>;        ■
```

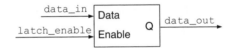

Figure 2–4: Latch Symbol

Listing 2–7. **Latch Design**

```
1.      Library IEEE;
2.      Use IEEE.std_logic_1164.All;
3.      Entity Latch Is Port (
4.         data_in            : In std_logic;
5.         latch_enable       : In std_logic;
6.         data_out           : Out std_logic);
7.      End Latch;
8.
9.      Architecture arch_Latch Of Latch Is
10.     Signal internal_data_out      : std_logic;
```

```
11.
12.     Begin
13.     -- creating a latch signal
14.         internal_data_out        <=  data_in  When  latch_enable  =  '1'
    Else internal_data_out;
15.
16.         data_out                 <= internal_data_out;  -- output signal is
    assigned internal signal value
17.     End arch_Latch;
```

Line 10. Internal Signal Created for Output Signal Data Assignment

This latch is created with a concurrent statement; therefore, it does not appear in a process. In VHDL, entity signals defined as outputs can only be assigned values and cannot be used for things like conditions or calculations. The reason this is important is because the output signal data_out needs to be used in the concurrent statement. Since this is not allowed in VHDL, the problem is solved by creating the internal signal internal_data_out.

Line 14. Latch Data

The internal data_out latch is created by the When statement.

Line 16. Output Signal Assigned Value of Internal Signal

The output signal is then set equal to the internal signal. The internal signals must have the same data type and size as the output signal.

2.7. Shift Register

Figure 2–5 shows a simple shift register, where a single bit is shifted from the LSB to the MSB. There are a couple of ways to write the code. Option 1 requires manually writing each bit shift signal assignment, as shown in Listing 2–8.

Lines 23–28

Each bit is assigned the value of the data bit to its right.

Figure 2–5: Shift Register

Listing 2–8. Manual Shift Register

```
1.     Library IEEE;
2.     Use IEEE.std_logic_1164.All;
3.
4.     Entity ShiftRegister Is Port (
5.       clk                  : In std_logic;
6.       reset                : In std_logic;      -- power-on reset
7.       shift_data           : In std_logic;
8.       shifted_data_out     : Out std_logic_vector (5 Downto 0));
9.     End ShiftRegister;
10.
11.    Architecture arch_ShiftRegister Of ShiftRegister Is
12.    Signal internal_shifted_data_out     : std_logic_vector (5 Downto 0);
13.
14.    Begin
15.    shifted_data_out      <= internal_shifted_data_out; -- output signal
   is assigned internal signal value
16.
17.    left_shift: Process (clk, reset)
18.    Begin
19.      If reset = '1' Then
20.        internal_shifted_data_out          <= ((Others => '0' ));
21.      Elsif rising_edge (clk) Then
22.        -- manually creating a shift register
23.        internal_shifted_data_out(0)         <= shift_data;
24.        internal_shifted_data_out(1)         <= internal_shifted_
   data_out(0);
25.        internal_shifted_data_out(2)         <= internal_shifted_
   data_out(1);
26.        internal_shifted_data_out(3)         <= internal_shifted_
   data_out(2);
27.        internal_shifted_data_out(4)         <= internal_shifted_
   data_out(3);
28.        internal_shifted_data_out(5)         <= internal_shifted_
   data_out(4);
29.      End If;
30.    End Process;
31.    End arch_ShiftRegister;
```

Option 1 is okay for smaller shift registers, but it can be time consuming for larger numbers of bits. Option 2 uses the reserved word downto to represent the individual shifts with fewer lines of code.

Listing 2–9. Simplified Shift Register

```
1.    shift_values: Process (clk, reset)
2.    Begin
3.       If reset = '1' Then
4.          shifted_data_out      <= ((Others => '0'));
5.       Elsif rising_edge (clk) Then
6.       -- this is the shortcut to creating a shift register
7.          internal_shifted_data_out(0)          <= shift_data;
8.          internal_shifted_data_out(5 downto 1) <= internal_shifted_data_out
   (4 downto 0);
9.       End if;
10.   End Process;
```

Lines 7–8. Shift Register Shortcut

Remember, in the signal assignment, the MSB number is written prior to downto with the LSB following. With that in mind, the signal assignments for internal_shift_data_out in the shift_values process can be rewritten using downto, as shown in Listing 2–9.

The downto signal assignment means that internal_shifted_data_out bit 1 is assigned the value of internal_shifted_data_out bit 0, internal_shifted_data_out bit 2 is assigned the value of internal_shifted_data_out bit 1, and so on. The shift register signal assignment can be written using one signal statement; however, the operator for that assignment is not discussed in this book.

2.8. Comparator

Relational operators such as greater than, >; greater than or equal to, >=; less than, <; less than or equal to, <=; and equal, = are used for comparisons. These operators accept std_logic and integer values and return a Boolean (true or false) value. Figure 2–6 shows a simple example where two numbers are compared.

Line 14. Compare Statement

A concurrent statement using When can be used to determine if a number is smaller than a second number. Whenever number 1 is smaller than number 2, the output num1_small_num2 goes high. The design code is shown in Listing 2–10.

Figure 2–6: Compare

Listing 2–10. Comparator

```
1.   Library IEEE;
2.   Use IEEE.std_logic_1164.All;
3.
4.   Entity Comparison Is Port (
5.      number_1                 : In std_logic_vector (2 Downto 0);
6.      number_2                 : In std_logic_vector (2 Downto 0);
7.      num1_smaller_num2        : Out std_logic);
8.   End Comparison;
9.
10.   Architecture arch_Comparison Of Comparison Is
11.
12.   Begin
13.   -- an example of a simple comparison; the output goes high when number 1
    is smaller than number 2
14.      num1_smaller_num2        <= '1' When number_1 < number_2 Else '0' ;
15.
16.   End arch_Comparison;
```

2.9. Binary Counter

The counter shown in Figure 2–7 is a 4-bit binary counter that automatically increments on each rising edge of the input clock, see line 22. The design code for this counter is shown in Listing 2–11.

A synchronous enable is easily added to this binary counter, as shown in Figure 2–8, with a change to the entity and the process in the architecture section.

The entity changes shown in Listing 2–12 adds line 4.

Line 4. Enable Input

The enable signal named is added to the entity.

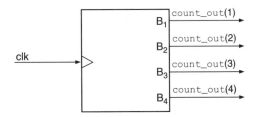

Figure 2–7: Binary Counter

Listing 2–11. Binary Counter

```
1.      Library IEEE;
2.      Use IEEE.std_logic_1164.All;
3.      Use IEEE.std_logic_signed.All;
4.
5.      Entity BinaryCounter Is Port (
6.          clk             : In std_logic;      -- master input clock
7.          reset           : In std_logic;      -- power-on reset
8.          count_out       : Out std_logic_vector (3 Downto 0));   -- output
    value from counter
9.      End BinaryCounter;
10.
11.     Architecture arch_BinaryCounter Of BinaryCounter Is
12.     Signal internal_count_out       : std_logic_vector (3 Downto 0);
13.
14.     Begin
15.         count_out      <= internal_count_out;      -- internal counter
    result being assigned to output signal
16.
17.     counter: Process (clk, reset)
18.     Begin
19.        If reset = '1' Then
20.            internal_count_out      <= (Others => ('0'));  -- resetting
    initial output value of counter
21.        Elsif Rising_Edge (clk) Then
22.            internal_count_out      <= internal_count_out + "0001";
           -- increment counter
23.        End If;
24.     End Process;
25.     End arch_BinaryCounter;
```

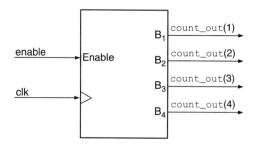

Figure 2–8: Binary Counter with an Enable

Listing 2–12. Entity Changes to Binary Counter with Synchronous Enable

```
1.    Entity SyncBinaryCounter Is Port (
2.      clk              : In std_logic;      -- master input clock
3.      reset            : In std_logic;      -- power-on reset
4.      enable           : In std_logic;
5.      count_out        : Out std_logic_vector (3 Downto 0));   -- output
   value from counter
6. End SyncBinaryCounter;
```

Listing 2–13. Process Changes to Binary Counter with Synchronous Enable

```
1.  synccounter: Process (clk, reset)
2.  Begin
3.    If reset = '1' Then
4.        internal_count_out        <= (Others => ('0'));
5.    Elsif rising_edge (clk) Then
6.      -- adding a synchronous enable to the binary counter
7.      If enable = '1' Then
8.          internal_count_out  <= internal_count_out + "0001";
      -- increment counter
9.      End If;
10.   End If;
11.   End Process;
```

The changes to the counter process in the architecture section, shown in Listing 2–13 adds lines 7 and 9.

Line 7. Synchronous Enable

Insert enable If-Then condition following the rising edge statement in the synccounter process.

2.10. Conversion Functions

The conversion functions allow you to convert from one data type to another. The conversion functions can be concurrent or sequential statements. There are many reasons why you may need or want to convert data types. You may not like working with a specific data type or the input data type may be unacceptable for the operation you want to perform. For example, all the input signals in this section are std_logic or std_logic_vector data types; however, to perform division using the numeric_std package requires the data type to be either signed, unsigned, natural, or integer, reference Table 2–2.

Table 2–2: Conversion functions

Conversion Function	Converted Data Type	Package
`to_integer`	Signed to integer or unsigned to natural	
`to_unsigned`	Natural to unsigned	`numeric_std`
`to_signed`	Integer to signed	
`to_stdlogicvector`	`bit_vector` or `std_ulogic_vector`	`std_logic_1164`
`to_integer`	Unsigned to natural or signed to integer	`numeric_bit`
`Conv_std_logic_vector`	Integer to standard logic vector, or unsigned to standard logic vector, or signed to std logic vector	`std_logic_arith`
`Conv_integer`	Signed or `Std_ulogic` to integer	
`Conv_integer`	`Std_logic_vector` to integer	`std_logic_unsigned`

Some conversion functions and their packages are listed in Table 2–2. Using the `conv_integer` function in the `std_logic_unsigned` package on the `std_logic` inputs makes it possible to perform the division operation.

Additional Library Packages Added

In Listing 2–14, line 3 adds the `std_logic_unsigned` package from the IEEE library, so the `conv_integer` operation could be used. Line 4 adds the `std_logic_arith` package from the IEEE library, so the `conv_std_logic_vector` operation could be used.

Listing 2–14. Conversion

```
1.    Library IEEE;
2.    Use IEEE.std_logic_1164.All;
3.    Use IEEE.std_logic_unsigned.All;
4.    Use IEEE.std_logic_arith.All;
5.
6.    Entity Convert2Integer Is Port (
7.      number_1          : In std_logic_vector (3 Downto 0);
8.      number_2          : In std_logic_vector (3 Downto 0);
9.      quotient          : Out std_logic_vector (3 Downto 0));
10.   End Convert2Integer;
11.
12.   Architecture arch_Convert2Integer Of Convert2Integer Is
13.   Signal integer_num1      : integer;
14.   Signal integer_num2      : integer;
15.   Signal integer_quotient  : integer;
16.
```

```
17.   Begin
18.   -- standard logic vector numbers are converted to integer before
   performing division
19.      integer_num1        <= Conv_integer (number_1);
20.      integer_num2        <= Conv_integer (number_2);
21.      integer_quotient    <= integer_num1 / integer_num2;
22.      quotient            <= Conv_std_logic_vector
   (integer_quotient,3);
23.
24.   End arch_Convert2Integer;
```

2.11. Read File

The read command extracts information from an external file, see Figure 2–9. Such data can be fed into input signal(s) defined by an entity. Some of the data types that can be read from the external file using the read command are integers, Boolean, character, time, real, and string. This command is very useful when verifying the design code and is discussed more in the simulation chapter.

The code in Listing 2–15 reads integer data from an external text file. These data are converted to standard logic vectors and used as input data to the design component named *readfile*. The functions of the readfile are not defined; however, the design file could process the input data and provide a single output resultant or decision bit.

Lines 4–5. Additional Library and Package Added

The read command is defined in the `std.textio` package in the std library. This library and package are made visible and usable to the program in the library declaration section.

Line 31. Defining File to Be Read

Before the data can be read from the file it must be defined using the reserved word `file`. This command states that the file provides data, it is opened in the read mode, and the name of the file is `read_data.txt`. The syntax for file is

File input: Text **Open** read_mode **Is** "Std_Input";

Where `input` is `data_in` and `std_input` is `read_data.txt`.

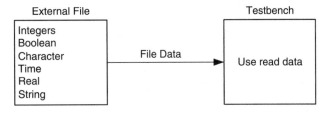

Figure 2-9: Read External File

Listing 2–15. Read **Command**

```
1.  Library IEEE;              -- define library and packages needed
    for this design
2.  Use IEEE.std_logic_1164.All;
3.  Use IEEE.std_logic_arith.All;
4.  Library std;              -- read command located in std_textio library
5.  Use std.textio.All;    -- needed to use file command
6.
7.  Entity testbench Is End testbench;
8.
9.  Architecture tb_read Of testbench Is
10.
11.     Signal clk              : std_logic := '0';
12.     Signal reset            : std_logic := '1';
13.     Signal input_data1      : std_logic := '0';
14.     Signal input_data2      : std_logic := '0';
15.     Signal input_data3      : std_logic := '0';
16.     Signal output_data      : std_logic;
17.     Signal data_vec         : std_logic_vector (2 downto 0);
18.
19.     Constant twenty_five_nsec    : time := 25 nsec;
20.
21.     Component readfile
22.     Port(
23.         clk                 : In std_logic;
24.         reset               : In std_logic;
25.         input_data1         : In std_logic;
26.         input_data2         : In std_logic;
27.         input_data3         : In std_logic;
28.         output_data         : Out std_logic);
29.     End Component;
30.
31.     File data_in: Text Open read_mode Is "read_data.txt";      -- define file
    that will be read
32.
33.     Begin
34.
35.     read_component: readfile
36.     Port Map(
37.         clk                 => clk,
38.         reset               => reset,
39.         input_data1         => input_data1,
40.         input_data2         => input_data2,
41.         input_data3         => input_data3,
42.         output_data         => output_data);
```

```
43.
44.    reset     <= '0' after 100.00 nsec;        -- set power-on reset inactive
45.
46.    create_twenty_Mhz: Process      -- create 20MHz simulation clock
47.    Begin
48.      Wait For twenty_five_nsec;
49.      clk       <= Not clk;
50.    End Process;
51.
52.    read_file: Process
53.    Variable data_line        : line;
54.    Variable data_integer     : integer;
55.
56.    Begin
57.      While Not endfile(data_in) Loop
58.        readline (data_in, data_line);
59.        read (data_line, data_integer);
60.        data_vec        <= conv_std_logic_vector(data_integer,3);
61.        input_data1    <= data_vec(0);
62.        input_data2    <= data_vec(1);
63.        input_data3    <= data_vec(2);
64.        wait for 25 nsec;
65.      End Loop;
66.      file_close(data_in);
67.    End Process;
68.    End tb_read;
```

Lines 53–54

Variables are used to define a line and an integer because they are needed to use the read and readline commands.

Lines 57–65

A While loop is used to read from the external data file.

Line 58

The first thing that must be done to get the data from the file is to read a line. This is done using the readline command. The syntax is

```
Readline (file f: Text; L: out Line);
```

where file `f: Text` is the file name; `data_in` is defined immediately following reserved word `file` on line 31; `L: out Line` is the variable `data_line` defined on line 53.

Line 59

Next, the first element on the line is read using the read command. The read command syntax is

```
read( L: inout line; Value: <certain data types>);
```

where `L: inout line` is the variable `data_line` defined on line 53. `Value` can be data types `bit_vector`, Boolean, character, integer, real, string, and time. In this case, it is an integer.

Lines 60–63

The data in the external file are integers. Once the data have been read, the integers are converted to a 3-bit standard logic vector. Each data bit is assigned to specific inputs, which are used to simulate or verify the design.

It is not always necessary to convert the data; this depends on the data type of the interface signal(s).

Line 66

The external file is closed when the loop is done.

2.12. Write File

The `write` command puts data to an external file, see Figure 2–10. Some of the data types that can be written to the external file using the `write` command are integer, Boolean, character, time, real, and string. This command can make verifying a design easier by having the results written to an external file. Data types that can be written are `bit`, `bit_vector`, Boolean, character, integer, real, string, and time. Data written to the file can be any of the data types or a combination. An example of the `write` command is shown in Listing 2–16.

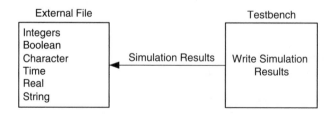

Figure 2–10: Write External File

Listing 2–16. Write **Command**

```
1.      Library IEEE; Use IEEE.std_logic_arith.All;
2.      Library std;
3.      Use std.textio.All;
4.
5.      Entity testbench Is End testbench;
6.
7.      Architecture tb_write Of testbench Is
8.
9.      Signal clk                     : std_logic := '0';
10.
11.     Constant twenty_five_nsec      : time := 25 nsec;
12.
13.     File data_out: Text Open write_mode Is "write_data.txt";
14.
15.     Begin
16.
17.     create_twenty_Mhz: Process
18.     Begin
19.       Wait For twenty_five_nsec;
20.       clk  <= NOT clk;
21.     End Process;
22.
23.     write_results: Process (clk)
24.     Variable data_line  : line;
25.
26.     Begin
27.       If rising_edge(clk) Then
28.         writeline (data_out, data_line);
29.         write (data_line, string'("Hello World."));
30.       End If;
31.     End Process;
32.     End tb_write;
```

Lines 2–3

The write command is defined in the `std.textio` package in the std library. This library and package is made visible and usable to the program in the library declaration section.

Line 13

The file command used to define the external data file is similar to the read command. The file syntax used for the write command is

```
File output: Text open write_mode Is "std_output";
```

where `output` is `data_output` and `std_output` is `write_data.text`.

My simulator puts the write file in the same directory as my work directory; this is discussed further in the simulation chapter. Consult the documentation for your simulation to determine where the file will be located.

Line 24

This is the defined line variable, `data_line`, to be used with `writeline` and `write` commands.

Line 28

The `writeline` command is used to write a line to the external file. Syntax for the `writeline` is

```
Writeline(file f: Text; L: inout Line);
```

where `file f: Text"` is `data_out`, defined on line 13, and `L: inout Line` is `write_data.txt`, defined on line 13 and is the file where the data is written.

Line 29

Hello World is written in the `write_data.txt`.

2.13. Chapter Overview

In this chapter, you were presented with some very simple design code and shown how the code can easily be modified to perform more advanced functions. As a part of the FPGA development phases, you will learn how to verify or simulate design code like the ones presented in this chapter. However, if you are interested in verifying some of the designs provided in this chapter, the book's appendix provides testbenches for all design code except `read` and `write` commands, which are testbenches.

Key Simple Design Points

- Start small and gradually build the design. It is much easier and less frustrating if you build a little and verify instead of trying to build and troubleshoot everything all at once.

- When writing code from a schematic design, make your signal names match.

CHAPTER 3

FPGA Development Phases

3.1. Introduction

The purpose of this chapter is to discuss the basic FPGA architecture and introduce the five FPGA development phases. It is important to understand something about the hardware aspect when developing an FPGA design. By *hardware*, I mean the FPGA device itself. Many of the older FPGAs allowed only one-time programming (OTP), which means that any design change required replacing the current device. However, most of today's FPGAs are based on static random access memory (SRAM) and can be reprogrammed multiple times. The development process requires the use of several tools to manipulate the design and produce an output file that dictates how the actual FPGA device is connected internally. This chapter presents a couple ways that FPGA manufacturers present their data. However, after reading this chapter, you should be able to apply this information to any manufacturer.

In this chapter, you will learn

- Basic FPGA architecture.

- Three basic FPGA capabilities.

- Altera and Xilinx FPGA architecture.

- Introduction to the five FPGA development phases.

3.2. What Is a Field Programmable Gate Array?

An FPGA is a device that consists of thousands or even millions of transistors connected to perform logic functions. They perform functions from simple addition and subtraction to complex digital filtering and error detection and correction. Aircraft, automobiles, radar, missiles, and computers are just some of the systems that use FPGAs.

A main benefit to using FPGAs is that design change(s) need not have an impact on the external hardware. Under certain circumstances, an FPGA design change can affect the external hardware (i.e., printed wiring board), but for the most part, this is not the case. One

Doi:10.1016/B978-1-85617-706-1.00003-5

situation would be if the device has insufficient resources to support the design changes, then a new device is required. If the new device is not a direct drop-in replacement—meaning pin-for-pin compatible (i.e., power and grounds are in the same location)—then the printed wiring board must be modified. More times than not, your design changes do not affect the hardware, especially if you have derated or left room in your device for growth. The amount of room for growth varies, but we talk more about this later in the book. The point I am trying to make is that FPGAs provide a lot of flexibility and opportunity to make design changes quickly.

Xilinx, Altera, and Quicklogic are just a few companies that manufacture FPGAs. Even though there are several FPGA manufacturers, they all share the same basic architecture concept. It consists of three basic capabilities: input/output (I/O) interfaces, basic building blocks, and interconnections. Figure 3–1 shows a generic FPGA architecture. It shows some basic building blocks connected to other basic building blocks, which are also connected to I/O interfaces, where data are passed to external sources. This figure is not meant to represent any particular device or design; it is provided only as a way of showing how the three basic capabilities interrelate. In the following sections, you are provided additional information on each of the capabilities.

3.3. I/O Interfaces

I/O interfaces are the mediums in which data are sent from internal logic to external sources and from which data are received from external sources. The interface signals can be unidirectional or bidirectional, single-ended or differential and could follow one of the different I/O standards. Some I/O standards are

- GTL (gunning transceiver logic).

- HSTL (high-speed transceiver logic).

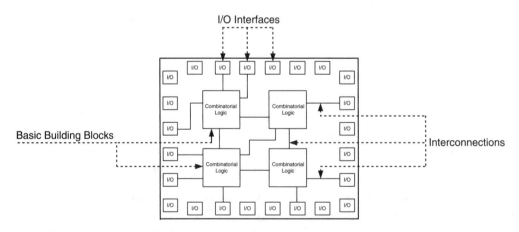

Figure 3–1: Generic FPGA Architecture

- LVCMOS (low-voltage CMOS).

- LVTTL (low-voltage transistor-transistor logic).

- PCI (peripheral component interconnect).

- LDT (lightning data transport).

- LVDS (low-voltage differential signaling).

The main purpose of the I/O interfaces is to transmit and receive data; however, the portion designated as an I/O interface may contain additional resources, such as voltage translators, registers, impedances, and buffers.

The term used for the I/O interface section may vary, depending on the manufacturer; however, the general function is the same. Consult the specific FPGA manufacturer's datasheet or application notes for the complete I/O interface details. As an example, a brief description of Altera's and Xilinx's I/O interface is presented.

Altera calls its I/O interfaces *I/O elements* (IOEs). They provide the basic internal to external interface function, support various differential and single-ended I/Os, and provide programmable pull-up resistors and I/O delays. The IOE structure for Cyclone II® is shown in Figure 3–2.

At Xilinx, the I/O interfaces are called *I/O blocks* (IOBs). The IOBs consist of registers, internal voltage translators, and other specialized resources. A simplified IOB diagram is shown in Figure 3–3.

As you can see, the Altera and Xilinx I/O interface sections use different terminology and have difference structures, but their basic function is the same, which is to pass I/O data between the device and external source(s). For complete I/O interface details, refer to the datasheet or application notes provided by the FPGA's manufacturer.

3.4. Basic Logic Building Blocks

The basic logic building blocks are preconfigured logic or resources used to build your design. What this means is that the FPGA starts with some basic logic, which is interconnected in various ways to perform the functions defined by the design. Altera's basic building blocks are called the *adaptive logic module* (ALM). The ALM consists of combinational logic, registers, and adders, see Figure 3–4. The combinational logic has eight inputs and a lookup table, LUT.

Xilinx's basic building blocks are called *configurable logic blocks* (CLBs). Each CLB contains slices, see Figure 3–5; and each slice has lookup tables (LUTs).

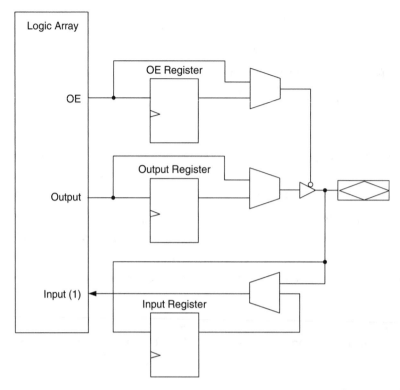

Figure 3–2: Cyclone II IOE Structure (This figure was reprinted with written permission from Altera Corporation. Altera is a trademark and service mark of Altera Corporation in the United States and other countries. Altera products are the intellectual property of Altera Corporation and are protected by copyright laws and one or more U.S. and foreign patents and patent applications.)

Each FPGA manufacturer defines the basic logic building block structure and the amount available. Because FPGAs are used in a variety of applications and systems, there is no "one device for all applications." FPGAs used in an aircraft might not be expected to perform the same types of functions or experience the same operating conditions as those in an automobile or personal computer.

There are many different types of FPGAs suitable for almost every kind of application. Selecting the right FPGA is made easier because they are divided into categories, often referred to as *families* or *series*. An FPGA family or series may have members or subfamily members. You can think of a family as a group of FPGAs with common characteristics that have members with distinctive features. The members share the basic family characteristics but have features that are distinctive from other family members, which may include the amount of memory, available resources, or number of I/O.

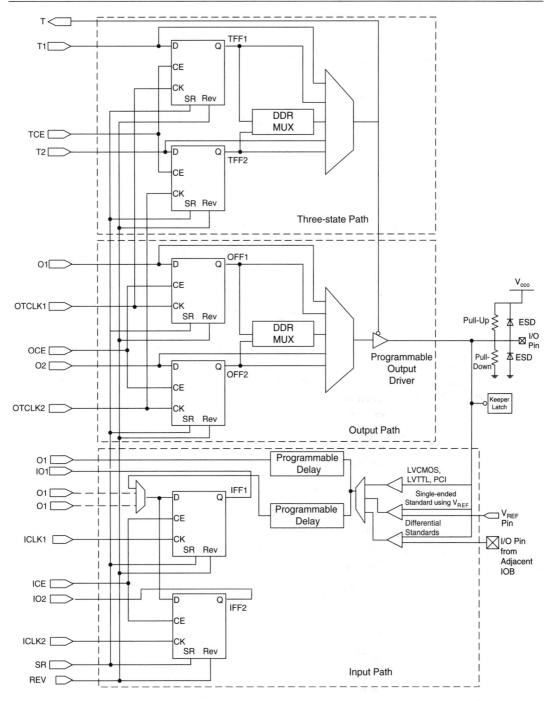

Figure 3–3: Simplified IOB Diagram (Material based on or adapted from figures and text owned by Xilinx, Inc., courtesy of Xilinx, Inc. Copyright © Xilinx. January 21, 2009.)

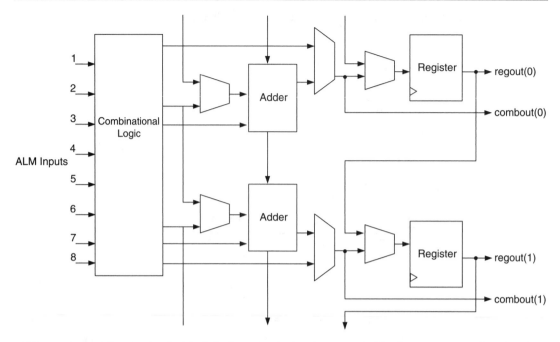

Figure 3–4: Adaptive Logic Module (ALM) Block Diagram1 (This figure was reprinted with written permission from Altera Corporation. Altera is a trademark and service mark of Altera Corporation in the United States and other countries. Altera products are the intellectual property of Altera Corporation and are protected by copyright laws and one or more U.S. and foreign patents and patent applications.)

Some FPGAs are characterized as having high volume, low cost, high temperature, or an embedded processor and are available in various sizes, packages, and speeds. Many manufacturers make device selection easier by grouping FPGAs according to their application (automotive, space, medical, etc.).

When you think about it, FPGA families are similar to a lot of our families, in that they have common characteristics, such as same last name and parents, but all the children are different in size, shape, personality, and unique in the way they work, act and think.

Manufacturers may categorize their devices differently, but do not let that throw you. Knowing some general things, like the intended system or application, should at least get to the right family or grouping. Before you buy, make sure you know what you are getting. Here is a brief overview of how Altera and Xilinx present their FPGA families and family members.

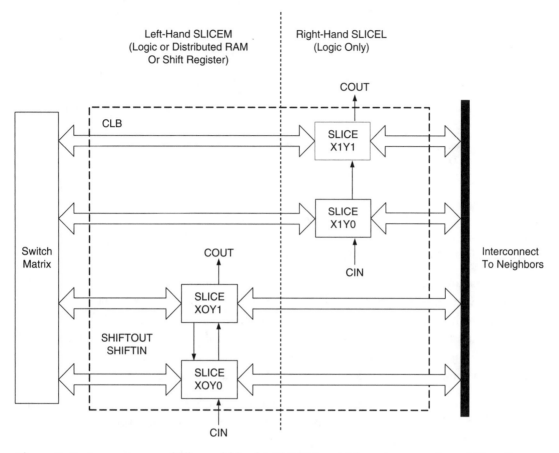

Figure 3–5: Arrangement of Slices within the CLB (Material based on or adapted from figures and text owned by Xilinx, Inc., courtesy of Xilinx, Inc. Copyright © Xilinx. June 25, 2008.)

Altera refers to its FPGAs in series. These series are

Stratix®

- High end and high density.

- On-chip transceivers.

Arria®

- Midrange.

- Transceiver based.

Cyclone®

- Low cost.

- Low power consumption.

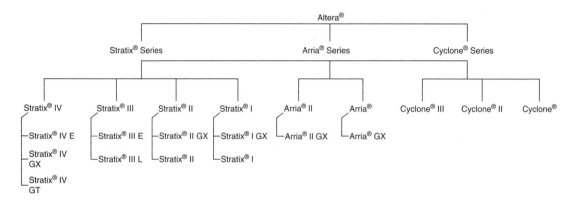

Figure 3–6: Altera Series Tree

Figure 3–6 shows the family members for these series.

Xilinx offers numerous FPGA families. Some of their families include Extended Spartan®-3A, Spartan-3E, Spartan 6, Virtex®, Virtex-E, Virtex-Pro, Virtex 5, and Virtex 6. These families are divided into members based on the amount of available resources. For example, Spartan-3E is a high-volume FPGA that has five family members, see Table 3–1. Note: The number following *S* in the member's name represents 1 for every 1000 system gates. This makes it easy to identify the number of systems gates just by looking at the part number.

I realize this can be a little confusing at first, but the more you work with it, the easier it becomes. Remember, the datasheets and application notes can be your best friend.

3.5. Ability to Interconnect

Interconnection involves connecting the basic building blocks to perform design-specific functions as well as connecting the internal logic to I/O interfaces, see Figure 3–7. Interconnection is performed automatically by the implementation tool, discussed in a later chapter. However, some tools allow the user to manually interconnect or route internal resources or logic. I recommend this only for advanced users.

Table 3–1: Spartan-3E family members data summary

Device	System Gates	Total CLBs	Total Slices	Max User I/O	Max Differential I/O Pairs
XC3S100E	100K	240	960	108	40
XC3S250E	250K	612	2,448	172	68
XC3S500E	500K	1164	4,656	232	92
XC3S1200E	1200K	2168	8,672	304	124
XC3S1600E	1600K	3688	14,752	376	156

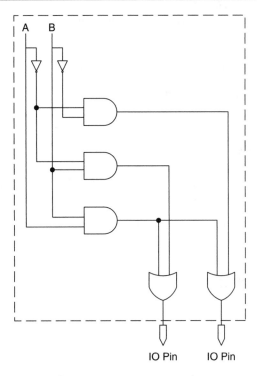

Figure 3–7: Interconnection

3.6. Programmable Logic Device Options

While this book focuses on basic FPGA design and features, they can be utilized in many different applications and perform advanced functions and calculations not discussed in this book. Some functions that were typically performed by a digital signal processor (DSP) are now being done by an FPGA, such as filtering and other signal processing. So, with that being said, I feel it is worth mentioning some other options available to designers.

The design approach used in this book is to manually write all the design code; however, this may not always be practical or provide the best result for the FPGA's design. What this means is that manually writing and verifying something like a finite impulse response (FIR) or fast Fourier transform (FFT) is time consuming. So, to make things a little easier, many FPGA manufacturers offer a variety of what are called *intellectual property* (IP) *cores* or *functions*. These IP cores or functions allow the designer to select and customize specific desired functions. The designer is generally presented with a graphical user interface (GUI), where information such as output format (VHDL, Verilog, etc.), target device (family, series, etc.), and the like are provided. The customized function provided by the tool can now be used in the design. Some of the advantages of using an IP core or function are

- Faster code development time.

- Reduced design risk, less likelihood of errors.

- Better and faster compiling.

- Better results for the design.

Some IP cores or functions are free, while others may be fee based. These IP cores or functions are manufacturer dependent.

Altera's IPs, called *Megafunctions*®, are designed for only their company's FPGAs. Figure 3–8 shows an example of Altera's IP Megafunction wizard. Here, the user selects the desired function and provides other information, such as the target family device and output format. The DSP option has been expanded to show the different available options.

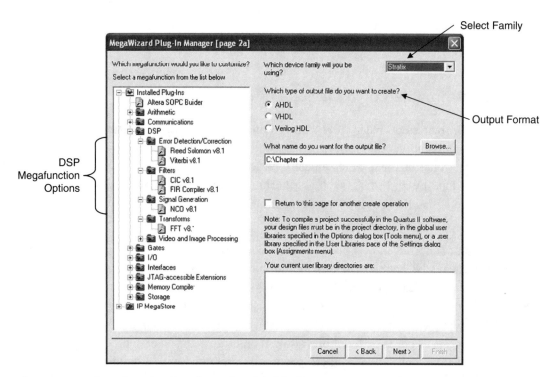

Figure 3–8: Altera IP Megafunction Wizard (This figure was reprinted with written permission from Altera Corporation.)

Xilinx's IPs are called *IP cores*. Some are offered Xilinx's free with their development tool, while others have to be purchased. Some of the IP cores offered with the company's ISE Software® development tool include

- Cascaded integrator comb, CIC, filter.

- FIR filter.

- FFT.

- Sine/cosine lookup table.

Here are some IP cores that must be purchased:

- Convolutional encoder.

- Reed-Solomon encoder.

- Reed-Solomon decoder.

- Controller area network (CAN).

Figure 3–9 shows an example of the GUI used to customize Xilinx's IP core. Additional information about the function, such as design output format (VHDL, Verilog), netlist format, or create a wrapper file, is added by selecting the part, generation, or advanced tabs located at the top of the screen.

3.7. FPGA Development Phases

Regardless of the design complexity, the FPGA development process is basically the same. For beginners or anyone who is trying to learn how to develop FPGA designs, the entire process can seem complex and confusing. What I found to be most confusing and hard to keep straight was all the terms, processes, and tools necessary to produce a design. When I first started with FPGAs, we used Synplicity's Synplify® for synthesis—talk about a mouthful. It seemed to take me forever to remember which was the company's name (Synplicity, now Synopsys), which was the tool's name (Synplify), and which was the process (synthesis).

I believe, when you are faced with trying to learn or perform complex tasks, things can be made easier when the tasks are divided into smaller pieces and tackled one at a time. This book takes the FPGA development process and divides it into five phases, which are discussed in the following chapters. The five FPGA development phases are design, synthesis, simulation, implementation, and programming, see Figure 3–10. A chapter is devoted to each development phase. Each chapter discusses the development phase's

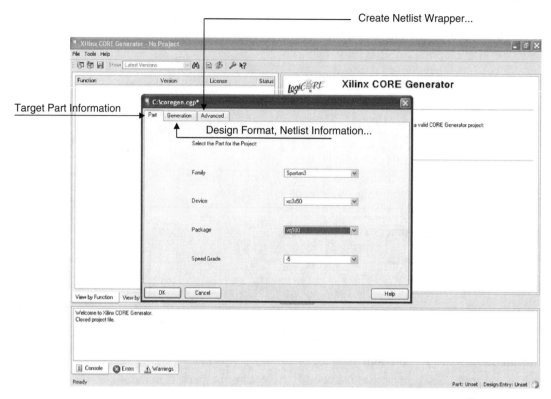

Figure 3-9: Xilinx's IP Core GUI (Material based on or adapted from figures and text owned by Xilinx, Inc., courtesy of Xilinx, Inc. Copyright © Xilinx 1995–2008 used in Xilinx ISE WebPack™ software version 10.1.)

inputs and outputs, tools, helpful Internet links, tips, and examples where appropriate. A tutorial is provided for the synthesis, simulation, implementation, and programming phases. The next chapter discusses the first FPGA development phase, which is design.

3.8. Chapter Overview

Over the years, FPGAs have come a long way. They can perform a wide range of operations from simple to complex. Initially, the development process seems complex and confusing, so it has been divided into smaller, less intimidating phases. Each of these phases can be as complex as the next, but learning the basic is a great way to start. Plus, after you have the basic, it is easy to expand your skills and knowledge.

Key Points to Remember

- The three basic FPGA architecture elements are I/O interfaces, basic building blocks, and interconnections.

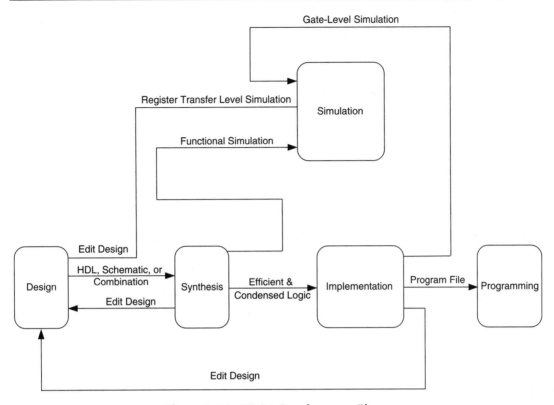

Figure 3–10: FPGA Development Phases

- Each manufacturer may present the basic FPGA architecture differently, but the concept is basically the same.

- Use the datasheets and application notes, because they provide helpful information, such as device resources and architecture structure, resource definitions and allocations, and design help.

- Do not try to tackle complex tasks all at once, reduce them into smaller, more manageable steps; before you know it, you have reached your goals.

Chapter Links

For your convenience, here is a list of links for more information on the IP cores and functions.

Xilinx's IP cores information can be found at www.xilinx.com/ipcenter/index.htm.

Altera's IP information can be found at www.altera.com/products/ip/getting-started/ipm-evaluate-download.html.

CHAPTER 4

Design

4.1. Introduction

FPGA development has been divided into five phases. The first development phase necessary to create the file that will be used to program an FPGA is design. This chapter discusses the types of material and other information that you may receive to create a design. The design package is information provided to a designer, and it can vary from design to design and company to company. With that in mind, this chapter presents some basic information that will help you understand the material, what information you need to get started, and how to develop a design from that information. The FPGA design may be as simple as converting a schematic to HDL or making modifications to an existing HDL design or more in depth, such as creating a totally "new" design. Regardless of the level of effect, I believe the approach presented in this chapter can partially or completely be applied to all types of designs. The final design from this phase is manipulated to produce the file that will program an FPGA. It is for that reason I consider this to be a critical phase. This chapter defines a systematic approach that will help you work through and understand the design phase.

In this chapter you will learn

- How to evaluate the design package.

- Decisions to make prior to creating the design.

- How to create a design.

4.2. What Is the Design Phase?

The design phase involves more than just creating a design. Many decisions must be made, and the design material must be understood. The "design package" is the input and the FPGA design is the output. After receiving the design package, the best way to start is to evaluate the design package and make some predesign decisions before writing the firmware, see Figure 4–1. The design code or code is also referred to as *firmware*. While HDL is

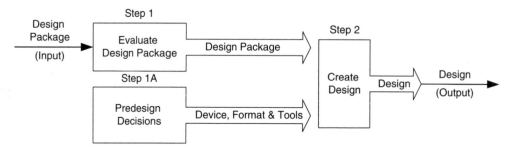

Figure 4–1: Design Phase Steps

softwarelike code, it is called *firmware* because it describes hardware and generally is written by hardware engineers. The definition of *firmware* can vary from person to person; some refer to languages such as C or C++ as *firmware*. For that reason, it is always a good idea to clarify the meaning. The design package contains the requirements that define the FGPA's features and functions. In other words, the requirements state what the design must do but not how to do it. During this phase, the design is created by writing firmware, through schematic capture, or a combination of the two. In this book, VHDL code is the design format. Success or failure of the design largely depends on:

- The quality of the design input(s):

  - A good design package is essential to creating the correct design.

  - You need to understand what needs to be designed and have the ability to create it.

- Making key decisions:

  - Selecting the FPGA for your specific application.

- Development tools:

  - A good text editor has features to help make the HDL design entry and modifications easier.
  - A graphical editor is needed to create and modify schematic capture designs.

4.3. Design Package

The design package is the result of predesign activities, generally performed by someone like a systems engineer, technical lead, or project engineer. These activities usually include:

- Creation of design architecture.

- Partitioning the design into sections.

  - The goal to minimize interfaces between sections and group together common functions.

- Designation of one or more designers.

  - Larger or complex designs may require several designers.

- Assign design sections to different designers, based on skill level or availability.

- Creation of design requirements.

  - Requirement specification defining what the design should do.

- Creation of Timing and other diagrams.

  - Diagrams provided as requirements or just as supporting documents.

The package should contain all the relevant design information, like requirements specification and timing diagrams. There should be enough information for a design to be created. It is never a good idea for the coder to create his or her own requirements. This leaves the door wide open for costly mistakes. Anything that is misinterpreted gets carried over into the design requirements, which become a part of the design. The design may be tested against the requirements and possibly delivered. This kind of mistake may not be detected until customer acceptance or, worse, after the design has been delivered, which can be very expensive and embarrassing. So my advice is this: If you are the coder, do not write your own requirements. The longer a mistake or error is carried into the development process, the more expensive it is to correct. Always remember, "Pay me now or pay me later; and if you pay me later, it will cost you more."

Design packages vary from project to project and company to company but should contain enough information to allow the designer or coder to create a final design. Some design packages have good documentation while others have vague or inconsistent information. I can hear my old coworkers laughing now, saying, "What design package or requirements?" We were often victims of "no formal requirements," "make it like the old," or "something verbal." I know a lot about bad design packages. Whether the design package is good or bad you should always evaluate its contents prior to starting the design. See Figure 4–2 for an example of a design package.

4.4. Evaluating the Design Package

Once you receive your design package, the urge is to immediately start creating the design. But wait, do not let your emotions take over. Some steps should be taken before you actually start the design. Taking this extra time in the beginning often reduces frustration and mistakes. It may feel like you are wasting time, but believe me it will be time well spent and oftentimes will result in faster progress. For some tips for evaluating your design package, see Figure 4–3.

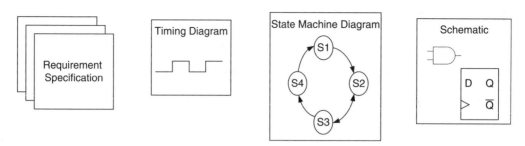

Figure 4–2: Design Package

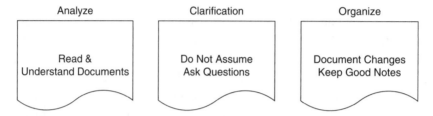

Figure 4–3: Design Package Evaluation Steps

4.4.1. Package Analysis

Read and understand all the documents provided in the design package. Be sure to have a clear understanding of what you are to design. Undoubtedly, as you read through the material, there will be questions and things that need to be clarified, so write down all your questions and get answers from the appropriate person(s). Question anything that seems to be contradictory or unclear.

4.4.2. Getting Clarification

In an ideal world, your design package will be crystal clear, have no contradictions, have everything in order, and be complete in every way. But, realistically, there may be some inconsistencies, pertinent information missing, outdated data, or just a variety of things that cause you to have questions. Whatever the case may be, do not be afraid to ask questions; and if you have doubt about what was presented, make no assumptions on the intent, because mistakes cost not only money but time. For answers, go directly to the source that gave you the material. Do not go to a coworker or friend for answers, because he or she may unintentionally give you incorrect information.

A word to the wise, make no assumptions or corrections without first trying to get clarification. What may appear to be a mistake or error may have a valid reason, and getting clarification can save you from having to redo work.

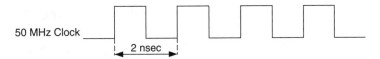

Figure 4–4: Inconsistent Documentation

For example, suppose you have a timing diagram that shows a 50 MHz or 20 nsec clock but the timing shows 50 MHz and 2 nsec, see Figure 4–4. There is no way of knowing which is correct, so do not assume. Instead, go back to the "source," not a coworker or friend, to get clarification and to clear up the inconsistencies.

I had a supervisor correct some of my work because he did not understand what was going on, only to discover that I was correct. After discovering what had occurred, the work had to be redone. This created confusion and wasted time and money, which could have been avoided if the supervisor had taken a little time to understand before acting on his assumptions or misunderstandings.

Additionally, if an acronym is provided without its definition, always verify the correct meaning, because, as you know, engineering has many acronyms and some have different meanings to different people.

While I caution you against making assumptions, the fact is that sometimes it may be necessary to make assumptions in order to make progress. If this is the case, then be sure to keep good records and document all assumptions.

4.4.3. Organization

Regardless of the number of documents in the design package, you should establish a system for storing the documents. The system should allow anyone to easily identify and retrieve the latest document revision. Because these documents are used to develop the design, it is very important to always make sure you work from the latest and most accurate information.

Ideally, the design package remains constant for the entire duration of the project; however, realistically, there is a good chance there will be a change that affects your design. So, if you don't already know you will soon find out that sometimes the only thing that remains the same is the "rate of change," so be prepared; because change may be necessary and may happen during any part of the development phases.

Sometimes design changes are informal, made during a meeting or verbally, and may not always get incorporated in the design package. If this is the case, make sure that your design package system allows you to create a paper trail or revision history for all changes,

■ Example 4–1. Documenting a Requirement Change

The original timing diagram shows a 50 MHz clock frequency with a conflicting 2 nsec time period. After verifying that 50 MHz is correct, 2 nsec is changed to 20 nsec, as shown in Figure 4–5, until the design package can be updated.

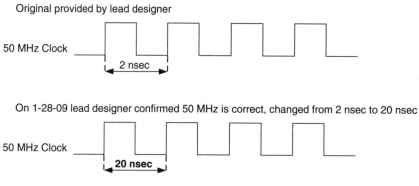

Figure 4–5: Corrected Timing Diagram

especially undocumented ones. At a minimum, I suggest that you document who requested the change, details about the change, and the date, similar to what is shown in Example 4–1. This provides a good record of all the design changes. Additionally, it provides traceability or a written record for yourself or, if necessary, a successor. Paper trails provide valuable information when things go wrong.

Now that you have completed your design package assessment, a few decisions must be made. These decisions are necessary because they affect various aspects of FPGA development.

4.5. Predesign Decisions

What is the design format? Who is the FPGA manufacturer? What tools should be used? These are some of the decisions that should be made prior to creating the design. However, if the design does not require manufacturer-dependent resources, such as memory, then you can wait until synthesis before deciding on the manufacturer and part number. Some options for each decision are shown in Figure 4–6. The design package may define one or some predesign decisions. For example, the requirements may state to modify a VHDL design to perform more functionality and put the new design in a larger FPGA package from the same manufacturer. In this case, the design format and manufacturer have been decided by the requirements; however, there are tool decisions that have to be made.

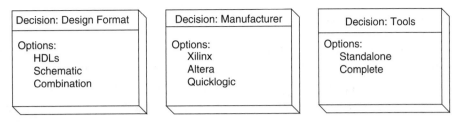

Figure 4–6: Design Decisions

On the other hand, making one decision can automatically determine the selection for another option. For example, selecting schematic capture allows you the freedom to select any manufacturer that accepts that design format. However, the tool is determined by the selected manufacturer. So selecting Altera for a schematic capture design means the Quartus II® design entry tool set is automatically selected.

While one decision can determine another, not all decisions must be made in the first development phase. Some decision can be made later. The design format must be selected in the first or design phase, since this is necessary to create the design. If the design is manufacturer independent, then the manufacturer and part number decisions can wait until synthesis. However, the manufacturer must be known in the design phase for manufacturer-dependent designs. Synthesis is the first phase in which the manufacturer and part number are needed for independent designs. This information is needed to set up the synthesis tool, so the correct output is generated for the implementation phase. Table 4–1 shows that the design format is selected in the design phase; the manufacturer is selected in design phase for manufacturer-dependent designs, otherwise it is selected during synthesis. The tools are generally selected during the specific phase unless they are predetermined by some other factors or a previous decision.

4.5.1. Design Format

Prior to creating the design, you must select the design's format. Will it be schematic capture, HDL, or a combination of the two? Sometimes, this decision has been made by your design package, as shown in Example 4–2.

Table 4–1: Decision/Development Phase Relationship

Decision	Design Format	Manufacturer	Tools
Design phase	X	X*	X*
Synthesis phase	–	X	X
Implementation phase	–	–	X
Programming phase	–	–	X

*Required for manufacturer-dependent designs.

■ Example 4–2. Predesign Decisions

You are required to convert a schematic capture design to VHDL. However, you can select the manufacturer and development tools. You may decide to use Xilinx as the manufacturer, and your company may have standalone tools like Synplify® for synthesis and ModelSim® for simulation. For implementation, you are required to use the manufacturer's tool. So, you are provided the design format but are free to select the manufacturer and tools, see Figure 4–7.

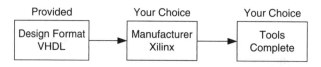

Figure 4–7: Example of Predesign Decisions

■

However, if you are starting a new design, then you may have the options to select one of the different design format options. As with anything there are advantages and disadvantages to each design format. For schematic capture,

Advantages:

- The design is drawn as a schematic.

- It is relatively easier to create, read, and understand.

Disadvantages:

- Logic symbols are proprietary, making the design manufacturer dependent.

- The entire design entry must be repeated for different manufacturers.

- Predefined logic symbols make the design less flexible.

- The options on development tools are limited.

For HDL,

Advantages:

- It has more design and manufacturer selection flexibility.

- To switch between different manufacturers, only manufacturer-specific resources, such as memory, or IP cores/functions have to be changed.

- Manufacturer-independent designs provide more development tool options.

Disadvantages:

- The design may be more difficult to read and understand.

- Manufacturer-dependent designs provide fewer development tool options.

4.5.2. FPGA Manufacturer

You may or may not have the option to select the actual FPGA. Sometimes, the manufacturer has been preselected, because of a company preference for a specific manufacturer, findings from a trade study, cost of the device, or a variety of reasons. Whether you have this option or not, knowing the FPGA's manufacturer is necessary for manufacturer-dependent designs. HDL designs that require no manufacturer-specific resources, such as memory, IP cores/functions allow you to create and simulate (verify) without knowing the manufacturer. However, for design entry, the manufacturer is needed for schematic capture and manufacturer-dependent HDL designs. Additionally, the manufacturer's part number is needed for the synthesis and place-and-route phases.

A little foresight is needed when selecting the actual device part number. This requires having an idea of how much resource your design requires. This can be difficult when you first start, but as you get more experience, you will be better able to determine which device best fits your needs. A good way to help you learn how to select devices is to randomly select a device, synthesize the design, and review the resources required in the output report. Now, you can see the resources required to perform the functions defined by your design. With this information you can use a datasheet to select a more appropriately sized device. Reading the synthesis report is a good way to understand what happens to the design as it get synthesized. Later in the book, you will learn more about synthesizing and its benefits.

Here are factors to consider when selecting the device:

Design application:

- Avionics

- Military

- Automotive

- Medical, and so forth.

Environmental conditions:

- Military.

- Industrial.

- Commercial.

Temperature range:

- Commercial, 0°C to 85°C.

- Industrial, –40°C to 100°C.

- Military, –55°C to 125°C.

Design size:

- Board allocated space.

- Package.

- Ball grid array, flat pack, and so on.

4.5.3. Development Tools

Each FPGA development phase utilizes specific tools, and they are discussed in the respective chapters. The design phase development tool depends mainly on the output format, but additional factors, such as cost, design sharing, and the need for it to be complete or standalone, can affect tool selection. For example, if your design format is schematic capture, then the design entry tool must be one that supports schematic capture and not a text editor. On the other hand, it may be easier to use complete development tools over standalone ones.

- **Design format**. For HDL, any text editor will work. See Section 1.5, "Editor Features," in Chapter 1, for tips on selecting a text editor. However, for schematic capture, you must use the tools supplied by the selected manufacturer.

- **Cost**. The fees for development tools can be very expensive, especially if they have yearly maintenance or licensing fees. Sometimes companies standardize the development tools, so you have no choice. But, if this is not the case, then I suggest having a clear understanding of your needs—try to get a temporary copy (i.e., try it before you buy it) and try to negotiate fees.

- **Design sharing**. For large designs that require multiple designers, your project may use tools that make it easy to divide the design among multiple people. So if this design has multiple coders, then you need a good set of tools to manage and control the design and its revisions.

- **Complete or standalone**. Some manufacturers offer complete development tools. When I say *complete development tool*, I mean that the tool provides design entry, synthesis, implementation, and simulation. For example, Xilinx's Integrated Software Environment (tm) (ISE) and Altera's Quantus II are complete development tools. Standalone tools perform a single function, such as synthesis, implementation, or simulation. For example, Synopsys's Synplify is used for design synthesis and Mentor Graphics's ModelSim simulator is used for design verification, which means neither of these tools can perform

Table 4–2: Complete tools, pros and cons

Pros	Cons
Single development tool	Can perform synthesis and implementation for only specific manufacturers
Manufacturer understands device better; therefore, development tool *may be* better and provide more accurate device data	Manufacturer is an expert on the device, not necessarily on the development tool
Single tool may be cheaper	Some development features may not be as good as a standalone tool

Table 4–3: Standalone tools, pros and cons

Pros	Cons
Supports multiple manufacturers	Must learn and use multiple development tools
Expert in specific tool area	Multiple cost or licensing fees
May have more or advanced features	Separate tools may be more expensive

the functions of the other. While it is true that some standalone tools, like ModelSim, provide a text editor that can be used to modify or create HDL code, the features and capabilities generally are not as good as standalone or dedicated text editors. I do not suggest using that type of editor for design entry.

Many pros and cons are involved in selecting complete development or standalone tools, see Tables 4–2 and 4–3.

As you can see, there are pros and cons to either solution. You have to determine which makes more sense for you and your application.

4.6. Creating Design Options

Now you are ready to create your design. FPGA designs can resemble a schematic, be written as HDL code, softwarelike language, or a mixture. Development options for schematic capture designs are limited. The symbols used to create the schematic are proprietary for the specific manufacturer, and development tools are available only from the specific manufacturer. Some HDL designs are manufacturer independent and can come in several formats. Two HDL development options are available: The design can be created using an automatic code generator or written manually. Each option is discussed in the following sections, but this book uses the manual option.

4.7. Automatic Code Generators

Automatic code generators provide an easy way to develop an HDL design without actually writing code. Such automatic code generators are a little different from the IP cores or functions discussed earlier. IP cores and functions produce code for only specific function(s), while the automatic code generator produces an entire design or major portions of the design. Automatic code generators work in different ways, depending on the software. Some software packages convert from one design format to another or convert a graphical model to HDL. For example, an ABEL design can be converted to VHDL. Several code generator tools create a templatelike skeleton VHDL testbench, the file used to verify the design (testbenches are discussed in a later chapter). For example, Doulos generates a testbench using a Perl script when provided with an entity or architecture. Automatic code generators are growing in popularity. Personally, I like to write the code. You will get a mixed bag of results from different automatic code generators. I am not sure that you will get all the comments you need or the format convention you want. While a lot of the work can be done by the auto code generator, you may still have to manually add code, comments, or formatting. I am not on the auto code generator bandwagon yet because I believe it never hurts to have the skills to develop firmware manually. As my mother always says "it is better to have it and not need it than to need it and not have it."

4.8. Manual Code Generation

VHDL, Verilog, and ABEL are some of the languages that can be used for FPGA design. Even though there are several language options, this book discusses the manual way to develop an FPGA design using VHDL. Manual code generation can be more time consuming than using an automatic code generator, but the designer has more control over the design. For those with little or no experience with HDL or programming, this chapter provides some helpful tips to consider as you review or create code. Regardless of the language you select, the FPGA development process is the same.

At this point you have a good understanding of the design requirements and what you are to design, assumptions have been documented, a manufacturer has been selected (and maybe the actual device part number), and you have your design entry tools. This is the point in the design phase where you use your design entry tool to create the design. I believe the best way to explain how to create a design is by example.

The design example is based on an information, friend or foe (IFF), system. IFF technology allows aircraft to be identified as a friend. The system consists of two major components: the interrogator, which requests information, and the transponder, which replies. Simply stated, the interrogator requests specific identification information from an aircraft by transmitting a series of pulses to the transponder. These pulses correspond to specific modes, which are decoded by the transponder and answered by transmitting a series of reply pulses. The entire system is complex, so for this example, the design is divided into a very small piece and the details are provided in a design package.

4.8.1. Design Package

From the requirements specification, you have determined that your assignment is to develop VHDL firmware that will receive two pulses and determine if the pulses have valid pulse width and spacing. Pulse spacing corresponds to two modes, Mode 2 and Mode 3A. The modes are decoded for valid pulses only, (i.e., with correct spacing). To ensure consistency among all the firmware, signal names and font conventions have been provided in the design package.

Font Conventions

* Capitalize the first letter of all reserved words.

* Lower case user-defined signals and nonreserved words with underscores, for easier readability.

Inputs

The P1 and P3 pulse width is 0.8 μsec + 0.1 μsec, see Figure 4–8.

VHDL signal name is `input_pulse`.

`Reset = active high`

VHDL signal name is `reset`.

`Master Clock = 20 MHz`

VHDL signal name is `clock20Mhz`.

Outputs

There are four output signals, defined as follows:

The narrow pulse signal is created when P1 or P3's pulse width is less than 0.7 μsec.

VHDL signal name is `narrow_pulse`.

The Wide Pulse signal is created when P1 or P3's pulse width is greater than 0.9 μsec.

VHDL signal name is `wide_pulse`.

Mode 2 decode is created when pulse spacing between valid P1 and P3 is 5 μsec ± 0.1 μsec, see Figure 4–9.

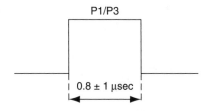

P1/P3

0.8 ± 1 μsec

Figure 4–8: P1 and P3 Pulse Width

Figure 4–9: Mode 2 Timing

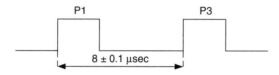

Figure 4–10: Mode 3A Timing

VHDL signal name is `mode2`.

Mode 3A decode is created when pulse spacing between valid P1 and P3 is 8 μsec ± 0.1 μsec, see Figure 4–10.

VHDL signal name is `mode3A`.

Evaluate

After reviewing your package everything seems clear, and there are no questions at this time.

Predesign Decisions

Design format was preselected as VHDL and my text editor is HDL Works' Scriptum 8.3 revision 1.

The manufacturer is Xilinx, no specific part number has been selected at this time.

For synthesis, Xilinx Synthesis Technology® (XST) synthesizer will be used. This is a part of Integrated Software Environment (ISE) Webpack®, which is a complete development tool. Also available is an evaluation copy of Synplify, so I decide to try both synthesis tools to see which gives better results. Although ISE Simulator® is a part of Xilinx's complete development tool, the standalone simulator ModelSim will be used for simulation. Implementation is performed using Xilinx's ISE complete development tool. The programmer will be decided during the implementation phase. This is acceptable, because information about the programmer is not needed until the end of implementation, when it is time to generate the programming file.

Design

Before starting the design, take a few moments to visualize how you will write the firmware. Will you use If-Then-Else statements, write a state machine, or what? To help me write the firmware, I created a state machine diagram, see Figure 4–11. Do not be afraid to create additional diagrams or other design aids to help you create the design.

Each coder has his or her design style; and as you gain more experience and learn different techniques, you will develop your own style. This design could have been written several ways; however, it reflects how I visualize the design, see Listing 4–1. The VHDL design consists of four main parts: an optional heading, library declaration, entity section, and architecture section.

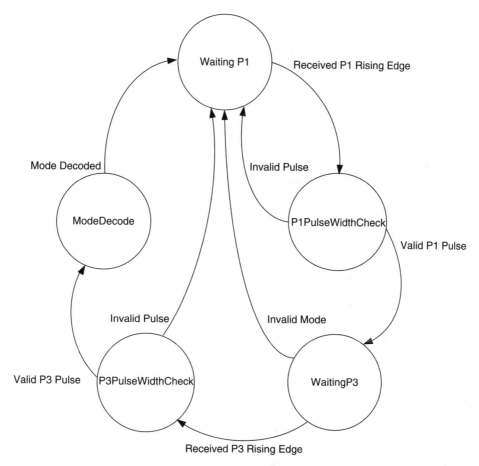

Figure 4–11: State Machine Diagram

Listing 4–1. Design Code

```vhdl
1.  --*************************** Header Section ***************************
2.  -- Name            :    Rebecca C. Smith
3.  -- Date            :    August 25, 2009
4.  -- Filename        :    mode2n3.vhd
5.  -- Description     :    This code performs pulse width and spacing checking.
6.  --                 :    For Pulse Width Checking:
7.  --                 :    Narrow pulse signal is sent for pulse widths less
                             than .7 µsec or
8.  --                 :    Wide pulse signal is sent for pulse widths greater
                             than .9 µsec.
9.  --                 :
10. --                 :    For Pulse Spacing (for valid P1 & P3 pulse widths
                             only):
11. --                 :    Mode 2 decode signal is sent for 5 µsec +/- .1 µsec
                             spacing
12. --                 :    Mode 3A decode signal is sent for 8 µsec +/- .1 µsec
                             spacing.
13. --
14. -- Revision History
15. -- Date                Initials            Description
16. --
17. --********************* End Header Section ************************
18. --
19. Library IEEE;           -- define library and packages needed for this design
20. Use IEEE.std_logic_1164.All;
21.
22. Entity mode2n3 Is Port (
23. clock20Mhz      :    In std_logic;      -- Master input clock
24. reset           :    In std_logic;      -- Power-on reset
25. input_pulse     :    In std_logic;      -- Input for P1 & P3
26. narrow_pulse    :    Out std_logic;     -- Indicates input pulse P1 or P3 is
                                               too narrow
27. wide_pulse      :    Out std_logic;     -- Indicates input pulse P1 or P3 is
                                               too wide
28. invalid_mode    :    Out std_logic;     -- Invalid P1 to P3 pulse spacing
29. valid_pulse     :    Out std_logic;     -- Indicates input pulse is valid
30. mode2           :    Out std_logic;     -- Valid Mode2 decoded
31. mode3A          :    Out std_logic);    -- Valid Mode3 decoded
32. End mode2n3;
33.
34. Architecture arch_mode2n3 Of mode2n3 Is
35.
36. Signal int_narrow_pulse  :  std_logic;
37. Signal int_wide_pulse    :  std_logic;
```

```
38. Signal int_invalid_mode      : std_logic;
39. Signal int_valid_pulse       : std_logic;
40. Signal int_mode2             : std_logic;
41. Signal int_mode3A            : std_logic;
42. Signal risingedge            : std_logic;    -- Indicates the rising
                                                    edge of P1 or P3
43. Signal fallingedge           : std_logic;    -- Indicates the falling
                                                    edge of P1 or P3
44. Signal sync_pulse            : std_logic;    -- Converts async signal
                                                    to sync
45.
46. Signal pulsewidth_counter    : integer;
47. Signal pulse_spacing         : integer;
48. -- used "type" to define state machine states
49. Type pulse_states Is (waitingP1, p1pulsewidth_check, waitingP3,
    p3pulsewidth_check, decode_mode);
50.
51. Signal current_state         : pulse_states;
52.
53. Begin
54. -- begin assigning internal signals to corresponding output signals
55. narrow_pulse           <= int_narrow_pulse;
56. wide_pulse             <= int_wide_pulse;
57. mode2                  <= int_mode2;
58. mode3A                 <= int_mode3A;
59. invalid_mode           <= int_invalid_mode;
60. valid_pulse            <= int_valid_pulse;
61. -- end assigning internal signals to corresponding output signals
62.
63. -- rising edge detection
64. risingedge               <= '1' When sync_pulse = '0' And input_pulse = '1'
                                 Else '0' ;
65.
66. -- falling edge detection
67. fallingedge              <= '1' When sync_pulse = '1' And input_pulse = '0'
                                 Else '0' ;
68.
69. cur_state: Process (current_state, clock20Mhz, reset)
70. Begin
71. If reset = '1' Then        -- assigning power-on or initial states values
72.     pulsewidth_counter   <= 0;
73.     pulse_spacing        <= 0;
74.     int_narrow_pulse     <= '0' ;
75.     int_wide_pulse       <= '0' ;
76.     int_invalid_mode     <= '0' ;
77.     int_valid_pulse      <= '0' ;
```

```
78.    int_mode2                  <= '0';
79.    int_mode3a                 <= '0';
80.    current_state              <= waitingP1;
81. Elsif rising_edge (clock20Mhz) Then
82.
83.    Case current_state Is
84.       When  waitingP1 =>            -- initial state waiting to receive
                                           first pulse P1
85.          pulsewidth_counter  <=  0;         -- all  signals  are  assigned
                                                   inactive state values
86.          pulse_spacing         <=  0;
87.          int_narrow_pulse      <= '0';
88.          int_wide_pulse        <= '0';
89.          int_invalid_mode      <= '0';
90.          int_valid_pulse       <= '0';
91.          int_mode2             <= '0';
92.          int_mode3a            <= '0';
93.          If risingedge = '1' Then    -- Received P1 rising edge
94.             pulsewidth_counter    <= pulsewidth_counter + 1;
                                          -- start pulse width counter
95.             pulse_spacing         <= pulse_spacing + 1;
                                          -- start P1-P3 pulse
                                             spacing counter
96.             int_narrow_pulse      <= '0';
97.             int_wide_pulse        <= '0';
98.             int_invalid_mode      <= '0';
99.             int_valid_pulse       <= '0';
100.            int_mode2             <= '0';
101.            int_mode3a            <= '0';
102.            current_state         <= p1pulsewidth_check;
                                          -- move to next state & wait for P1
                                             falling edge
103.         Else
104.            current_state         <= waitingp1;
105.         End If;
106.
107.      When p1pulsewidth_check =>
108.         If pulsewidth_counter = 19 Then                  -- P1 is wide
109.            pulsewidth_counter    <= pulsewidth_counter;
110.            pulse_spacing         <= pulse_spacing;
111.            int_narrow_pulse      <= '0';
112.            int_wide_pulse        <= '1';    -- Send out wide pulse signal
113.            int_invalid_mode      <= '0';
114.            int_valid_pulse       <= '0';
115.            int_mode2             <= '0';
116.            int_mode3a            <= '0';
```

```
117.              current_state          <=  waitingP1;    -- Return  to  initial
                                                              state and wait for P1
118.         Elsif fallingedge = '1' Then                 -- Received P1 falling
                                                              edge
119.          If pulsewidth_counter   <= 13 Then          -- Pulse is narrow,
                                                              stop counters
120.             pulsewidth_counter   <= pulsewidth_counter;
121.             pulse_spacing        <= pulse_spacing;
122.             int_narrow_pulse     <= '1';             -- Narrow pulse signal
                                                              is active
123.             int_wide_pulse       <= '0';
124.             int_invalid_mode     <= '0';
125.             int_valid_pulse      <= '0';
126.             int_mode2            <= '0';
127.             int_mode3a           <= '0';
128.             current_state        <= waitingp1;    -- Return to initial
                                                              state and wait for P1
129.         Elsif pulsewidth_counter >= 14 And pulsewidth_counter <= 18
             Then -- valid pulse
130.             pulsewidth_counter   <= pulsewidth_counter;
131.             pulse_spacing        <= pulse_spacing + 1;
132.             int_narrow_pulse     <= '0';
133.             int_wide_pulse       <= '0';
134.             int_invalid_mode     <= '0';
135.             int_valid_pulse      <= '1';              -- P1 is good, activate
                                                              valid pulse signal
136.             int_mode2            <= '0';
137.             int_mode3a           <= '0';
138.             current_state        <= waitingp3;
139.          End If;
140.       Else    -- no falling edge or wide pulse, continue counting
141.          pulsewidth_counter   <= pulsewidth_counter + 1;
142.          pulse_spacing        <= pulse_spacing + 1;
143.          int_narrow_pulse     <= int_narrow_pulse;
144.          int_wide_pulse       <= int_wide_pulse;
145.          int_invalid_mode     <= '0';
146.          int_valid_pulse      <= int_valid_pulse;
147.          int_mode2            <= '0';
148.          int_mode3a           <= '0';
149.          current_state        <= p1pulsewidth_check;
150.       End If;
151.
152.    When waitingP3 =>
153.       pulsewidth_counter   <= 0;
154.       pulse_spacing   <= pulse_spacing + 1;  -- continuing counting
                                                     P1-P3 pulse spacing
```

```
155.          int_narrow_pulse      <= '0';
156.          int_wide_pulse        <= '0';
157.          int_invalid_mode      <= '0';
158.          int_valid_pulse       <= '0';
159.          int_mode2             <= '0';
160.          int_mode3a            <= '0';
161.      If pulse_spacing = 163 Then       -- P1-P3 spacing too wide for
                                                Mode2 or Mode3
162.          pulsewidth_counter    <= 0;
163.          pulse_spacing         <= 0;       -- stop counting
164.          int_narrow_pulse      <= '0';
165.          int_wide_pulse        <= '0';
166.          int_invalid_mode      <= '1';
167.          int_valid_pulse       <= '0';
168.          int_mode2             <= '0';
169.          int_mode3a            <= '0';
170.          current_state         <= waitingp1;      -- wait for
                                                          interrogation
171.      Elsif risingedge = '1' Then           -- Received rising
                                                   edge of P3
172.          If pulse_spacing <= 97 Or (pulse_spacing >= 104 And
              pulse_spacing <= 156) Then
173. -- P1-P3 outside Mode2 or Mode3 range
174.              pulsewidth_counter    <= 0;
175.              pulse_spacing         <= 0;            -- stop counting
176.              int_narrow_pulse      <= '0';
177.              int_wide_pulse        <= '0';
178.              int_invalid_mode      <= '1';
179.              int_valid_pulse       <= '0';
180.              int_mode2             <= '0';
181.              int_mode3a            <= '0';
182.              current_state         <= waitingp1;      -- wait for
                                                              interrogation
183.          Else
184.              pulsewidth_counter    <= pulsewidth_counter + 1;
                                                  -- start pulse width counter
185.              pulse_spacing         <= pulse_spacing;
                                                  -- stop pulse
                                                     spacing counter
186.              int_narrow_pulse      <= '0';
187.              int_wide_pulse        <= '0';
188.              int_invalid_mode      <= '0';
189.              int_valid_pulse       <= '0';
190.              int_mode2             <= '0';
191.              int_mode3a            <= '0';
192.              current_state         <= p3pulsewidth_check;
193.          End If;
```

```
194.          Else
195.             current_state          <= waitingP3;
196.          End If;
197.
198.      When p3pulsewidth_check =>
199.          If pulsewidth_counter = 19 Then          -- wide pulse
200.             pulsewidth_counter  <= pulsewidth_counter;
201.             pulse_spacing       <= pulse_spacing;
202.             int_narrow_pulse    <= '0';
203.             int_wide_pulse      <= '1';          -- P3 is wide send out
                                                       wide pulse signal
204.             int_invalid_mode    <= '0';
205.             int_valid_pulse     <= '0';
206.             int_mode2           <= '0';
207.             int_mode3a          <= '0';
208.             current_state       <= waitingP1;  -- Return to waiting
                                                       for rising edge of P1
209.          Elsif fallingedge = '1' Then          -- detecting P3 falling
                                                       edge
210.             If pulsewidth_counter <= 13 Then          -- narrow pulse
211.                pulsewidth_counter  <= pulsewidth_counter;
212.                pulse_spacing       <= pulse_spacing;
213.                int_narrow_pulse    <= '1';          -- send out narrow
                                                           pulse signal
214.                int_wide_pulse      <= '0';
215.                int_invalid_mode    <= '0';
216.                int_valid_pulse     <= '0';
217.                int_mode2           <= '0';
218.                int_mode3a          <= '0';
219.                current_state       <= waitingp1;  -- Return to waiting
                                                           for rising edge of P1
220.             Elsif pulsewidth_counter >= 14 And pulsewidth_counter <= 18
                 Then          -- valid pulse
221.                pulsewidth_counter  <= pulsewidth_counter;
222.                pulse_spacing       <= pulse_spacing;
223.                int_narrow_pulse    <= '0';
224.                int_wide_pulse      <= '0';
225.                int_invalid_mode    <= '0';
226.                int_valid_pulse     <= '1';          -- P3 is good, activate
                                                           valid pulse signal
227.                int_mode2           <= '0';
228.                int_mode3a          <= '0';
229.                current_state       <= decode_mode;
230.             End If;
231.          Else          -- no fallinge edge
232.             pulsewidth_counter  <= pulsewidth_counter + 1;
                                        -- continue pulse width counting
```

```
233.              pulse_spacing            <= pulse_spacing;
234.              int_narrow_pulse         <= int_narrow_pulse;
235.              int_wide_pulse           <= int_wide_pulse;
236.              int_invalid_mode         <= int_invalid_mode;
237.              int_valid_pulse          <= int_valid_pulse;
238.              int_mode2                <= '0';
239.              int_mode3a               <= '0';
240.              current_state            <= p3pulsewidth_check;
241.          End If;
242.
243.       When decode_mode =>
244.          If (pulse_spacing >= 98 And pulse_spacing <= 102) Then   --pulse
             spacing between 4.9 and 5.1 us
245.              pulsewidth_counter       <= pulsewidth_counter;
246.              pulse_spacing            <= pulse_spacing;
247.              int_narrow_pulse         <= int_narrow_pulse;
248.              int_wide_pulse           <= int_wide_pulse;
249.              int_invalid_mode         <= int_invalid_mode;
250.              int_valid_pulse          <= int_valid_pulse;
251.              int_mode2                <= '1';
252.              int_mode3a               <= '0';
253.              current_state            <= waitingP1;
254.          Elsif (pulse_spacing >= 158 And pulse_spacing <= 162) Then
             --pulse spacing between 7.9 and 8.2 us
255.              pulsewidth_counter       <= pulsewidth_counter;
256.              pulse_spacing            <= pulse_spacing;
257.              int_narrow_pulse         <= int_narrow_pulse;
258.              int_wide_pulse           <= int_wide_pulse;
259.              int_invalid_mode         <= int_invalid_mode;
260.              int_valid_pulse          <= int_valid_pulse;
261.              int_mode2                <= '0';
262.              int_mode3a               <= '1';
263.              current_state            <= waitingP1;
264.          Else
265.              pulsewidth_counter       <= pulsewidth_counter;
266.              pulse_spacing            <= pulse_spacing;
267.              int_narrow_pulse         <= int_narrow_pulse;
268.              int_wide_pulse           <= int_wide_pulse;
269.              int_invalid_mode         <= int_invalid_mode;
270.              int_valid_pulse          <= int_valid_pulse;
271.              int_mode2                <= '0';
272.              int_mode3A               <= '0';
273.              current_state            <= waitingp1;
274.          End If
275.
```

```
276.          When Others =>              -- if unknown state the signal will take on
                                             these values
277.              pulsewidth_counter     <= 0;
278.              pulse_spacing          <= 0;
279.              int_narrow_pulse       <= '0';
280.              int_wide_pulse         <= '0';
281.              int_invalid_mode       <= '0';
282.              int_valid_pulse        <= '0';
283.              int_mode2              <= '0';
284.              int_mode3A             <= '0';
285.              current_state          <= waitingp1;
286.      End Case;
287. End If;
288. End Process;
289.
290. edge_detect: Process (reset, clock20Mhz)
291. --This process syncs input pulse to master clock. This signal is used for
     edge detection.
292. Begin
293. If reset = '1' Then
294.     sync_pulse              <= '0';
295. Elsif rising_edge(clock20Mhz) Then
296.     sync_pulse              <= input_pulse;        -- input sync to master
                                                           clock
297. End If;
298. End Process;
299. End arch_mode2n3;
```

Part 1, Lines 1–17. Optional Heading Section

The original coder's name, original date, a brief description, and a revision history provide useful information for the reader.

Part 2, Lines 19–20. Library Declaration

Only the standard logic 1164 library is needed for this design. Its contents are made visible so they can be used in the design.

Part 3, Lines 22–32. Entity Section

Each input and output stated in the design package is listed and defined as standard logic.

Part 4, Lines 34–299. Architecture Section

The main design code is in the architecture section, which consists of processes, concurrent statements, and signal assignments.

There are two processes, which are found on lines 69–288 `Cur_state` process. This contains a case statement that is used to measure pulse width and pulse spacing and determine the mode.

Lines 290–298, `Edge_detect` process, are used to create a delayed copy of the input pulse for edge detection.

Lines 36–51, internal signal definitions, define internal signals used in the design.

Lines 46–47, counter signals defined, define the signals used for the pulse width and spacing counters.

Line 49, defining data type, creates a data type called `pulse_states` that can take on five values: `waitingp1`, `p1pulsewidth_check`, `waitingp3`, `p3pulsewidth_check`, and `decode_mode`.

In line 51, signal set to user-defined type, signal `current_state` is assigned `pulse_states` as the data type, which has five possible states.

In lines 55–60, output signals assigned internal signal data, concurrent statements are used to assign the values of the internal signals to the corresponding output signal.

The concurrent statement in lines 63–67, edge detection, created the rising edge detection that is needed.

Now that the VHDL design has been created there are two options as you move forward in the development process.

Option 1. Simulation

Simulation allows you to verify that the design meets requirements. No timing information is known, but design and logic errors can be found and corrected.

Option 2. Synthesis

Synthesis is the process that reduces the design and connects FPGA resources to perform the desired functions. While this process is required, it provides no way to determine if the firmware is performing the required functions. An optional simulation file can be provided; however, if error(s) are detected, it is difficult to determine if they are a result of the synthesis process or the design code.

Personally, I like to perform simulation next. This allows me to verify my design and make modifications as necessary. So, in the following chapter, the design is simulated.

4.9. Chapter Overview

The approach presented in this chapter should help you get started regardless of how much or little detail is provided in your design package. Remember, this is a very important phase, mistakes made in this stage get carried over into the other phases. The longer a mistake goes undetected, the more expensive it is to correct. Take time to produce your design; it will save you time and many headaches in the end.

Key Design Phase Tips

- Make sure that you evaluate your design package and get clarification when necessary.

- Develop a system to keep your documents organized.

- Remember, the longer errors or mistakes are undetected and carried further into the development phases, the more expensive and time consuming they are to correct.

Chapter Links

HDL Works Scriptum: www.translogiccorp.com/index.html.

Doulos's code generator: www.doulos.com/knowhow/perl/testbench_creation.

CHAPTER 5

Simulation

5.1. Introduction

This chapter discusses the simulation phase of the FPGA development process. In my opinion, it is the most exciting and fun part of the FPGA development phases. At this point, you have a design that needs to be verified. The design could be one created in this book, one you or someone else created, a netlist created as a result from the synthesis or implementation phases (discussed later in this book), or some modifications to an existing design. Regardless of the type of design, the purpose of simulation is to verify that the design performs the required functions. Typically, simulation can be performed in three places in the development process: on the output from design, synthesis, and implementation phases. Most often, simulation is performed only on the design and not on the netlist or file produced by synthesis or implementation phases.

Design verification can also be performed using lab equipment, such as logic analyzers; however, this can be a more time-consuming and a less effective way, especially for new designs. Lab verification is less flexible, in that errors can damage the hardware. Generally, the setup to re-simulate a design months later is easier and faster than obtaining and re-setting up lab equipment. Although simulation is not the only method to verify a FPGA design, I believe it to be the most feasible.

Even though simulation is not required, it should never be completely omitted from the FPGA development phase, especially for new designs. It may seem unnecessary to simulate small, simple changes, but those are the ones that can cause you the biggest problems. You can learn a lot about the design through simulation and have some fun in the process.

In this chapter, you will learn

- What tools are used for simulation.

- How to verify a VHDL design by writing testbenches.

- The options for collecting lab data for design verification.

5.2. What Is Simulation?

Simulation is the process of applying stimulus or inputs that mimic actual data to the design and observing the output. It is used to verify that the design performs the expected and required functions.

Inputs to the simulation phase can be the design phase output, synthesis netlist, and implementation netlist. Any one or all of these inputs can be used to perform simulation. The output can be a listing, where the data are represented as binary, hex, or the like; graphically as a waveform; or the final results (such as pass/fail indicator), see Figure 5–1. Output from simulation is unique in that it does not feed into another development phase. However, the output is very important, because it provides the medium that allows the tester or verifier to see how the design performs.

This phase is as important as the design phase, and as a general rule, the amount of time spent simulating should be about twice the design time. Ideally, the firmware should not be tested by the person who wrote the code. The original coder should do minimal testing, but comprehensive testing should be done by a code tester. The same reasoning why the original coder should not write his or her requirements apply to this situation. In addition to that reasoning, oftentimes the coder does not try as hard to find errors in his or her firmware as a

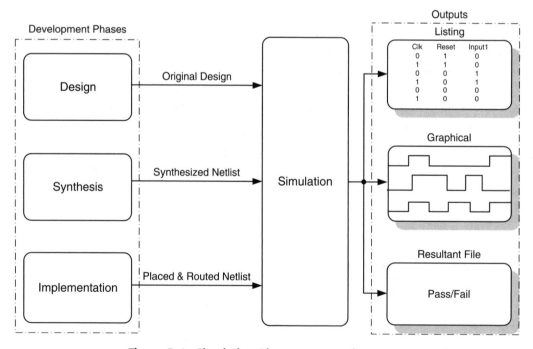

Figure 5–1: Simulation Phase Inputs and Outputs

third party. So the original coder may overlook or miss errors, such as design flaws that would be found by someone else. Once the design is complete, you must now verify that it performs as required. One way this is done is by performing a simulation on the design.

5.3. Simulation Tools

The tools needed in the simulation phase are an editor or editors and a simulator. The editor is used to create the inputs that will be applied to the design. As you simulate, it may be necessary to modify or change the firmware; therefore, you need an editor to modify the code created during the design phase. These editors may or may not be the same, depending on the format of each design. If you decide to create an HDL testbench (discussed later in this chapter), then a text editor is needed. However, if you are creating the inputs as a waveform, then you need a graphical editor. HDL code is generally supported by most, if not all, simulators; however, waveform test inputs may not be, so consult the documentation for your simulator. There are pros and cons for using HDL versus graphical editors for testing; many of these are the same as discussed in the design phase. Basically, HDL provides more flexibility, while waveforms are less flexible and not supported by all simulators.

The term *simulator* has been mentioned several times but not really explained. It is a tool that compiles or connects the test inputs to the design. Running the simulator feeds the input test data into the design, causing the outputs to change based on the input data, see Figure 5–2. The output data can be presented in several formats, such as a waveform, text file, or data formats (i.e., binary, hex, and so forth). There are many standalone simulators, such as Mentor Graphic's ModelSim®, which is my preference. Simulators can be very expensive, so I suggest doing a Web search. You should be able to find some free or trial offers for simulators that may meet your needs. Some manufacturers, like Xilinx and Altera, offer their own simulator brand with their complete package development tools. In addition to their

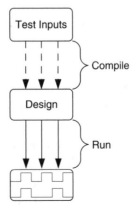

Figure 5–2: Design and Test Input Compile Flow

fee-based complete package development tools; they offer a free Web version of their design tools.

Here are some free Web downloads development tools with simulators:

Manufacturer:	Xilinx
Development tool:	ISE Webpack™, includes ModelSim XE III 6.4b and the "light" version of the ISE Simulator (ISim)
Download address:	www.xilinx.com/tools/designtools.htm

Note: The download may require a user name and login, which are free.

Manufacturer:	Altera
Development tool:	Quartus® II, Web Edition, offers ModelSim, Altera Starter Edition
Download address:	www.altera.com, then go to "Products" → "Design Software"

Note: The Quartus II, Web Edition, does not require a license.

The free tools offer fewer features and support fewer devices or operating systems than their fee-based counterparts, but these differences may not be an issue for you. Because some companies now offer free Web-based development tool packages, it is easy for anyone to download and learn a new skill or enhance his or her skills outside of work or even do home projects. This is a luxury that was not available when I started working with FPGAs.

5.4. Levels of Simulation

There are three levels of simulation, see Figure 5–3: the register transfer level (RTL), gate level, and functional level. Each level of simulation verifies different aspects of the design.

RTL performs simulation on the design phase code. Doing this prior to synthesizing allows you to troubleshoot the design for logic and syntax errors. The RTL simulation contains no timing information.

Functional simulation is performed on the netlist or the code generated by the synthesis tool. Oftentimes, it is necessary to direct the synthesis tool to generate the functional simulation netlist. Consult your synthesis tool's user's manual to determine if the netlist is generated automatically or is a selected option. The synthesized netlist allows you to verify that the synthesis process did not change the design. If you are going to perform a functional simulation, then a new netlist must be created each time the design is synthesized. The synthesis tool predicts and inserts some timing information, these are not the final timing delays. This simulation is more realistic than the RTL, but not as accurate when it comes to timing as the gate-level simulation.

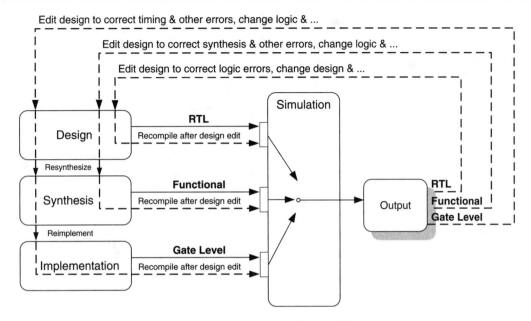

Figure 5–3: Simulation Levels

Gate-level simulation is performed on the code or netlist generated by the implementation tool. It may be necessary to direct the implementation tool to generate the gate-level simulation netlist. Consult your implementation tool's user's manual to determine if the netlist is created automatically or is a selected option. This simulation contains actual timing information and is the most realistic representation of the FPGA design. Now, the actual internal chip delays are known and incorporated into the netlist that represents the design. Because the actual timing of signals is known, timing problems can be detected during this simulation.

The ModelSim XE III 6.4b simulator included in Xilinx's ISE Webpack is used for the simulations in this chapter, unless otherwise indicated.

5.5. Test Cases

Test cases should be written prior to creating your testbench or graphical stimulus. Table 5–1 shows some test cases for the design. Test cases are written such that they verify that the design meets the requirements. No hard and fast rule states that you should have a specific number of test cases, but as a guide, there should be enough cases to verify the design. I have found that, more times than not, the smallest area you omit will come back to haunt you. So, try to make your test cases as complete as possible. However, for some designs, it may not be possible to test every possible situation, but do the best you can to ensure at least all the critical areas are covered.

The test cases created for the design code in this book are shown in Table 5–1. These test cases will be used to verify the design. The requirements state that the input pulses and modes must

Table 5–1: Design Test Cases.

Test Case	Mode 3 Pulse Spacing	Mode 2 Pulse Spacing	P3 Pulse Width	P1 Pulse Width	Expected Output
1	N/A	N/A	No pulse	Narrow	No decode
2	N/A	Normal	Narrow	Minimum	No decode
3	N/A	Normal	Minimum	Normal	Mode 2 decoded
4	N/A	Minimum	Normal	Maximum	Mode 2 decoded
5	N/A	–	Don't care	Wide	No decode
6	N/A	Maximum	Maximum	Normal	Mode 2 decoded
7	N/A	Normal	Wide	Normal	No decode
8	N/A	Narrow	Normal	Normal	No decode
9	N/A	Wide	Normal	Normal	No decode
10	Narrow	N/A	Normal	Normal	No decode
11	Minimum	N/A	Normal	Normal	Mode 3 decoded
12	Normal	N/A	Normal	Normal	Mode 3 decoded
13	Maximum	N/A	Normal	Normal	Mode 3 decoded
14	Wide	N/A	Normal	Normal	No decode

Note: *Narrow* is defined as any value less than the minimum, and *wide* is any value greater than the maximum.

meet a specific range for decoding to occur. Therefore, test cases are created to check P1 and P3 pulse widths that are below, at minimum, at normal, at maximum, and above the required range. Modes 2 and 3 pulse spacing is checked at minimum, normal, maximum, and above the required range. The code that checks the P1 and P3 pulse widths is the same code but copied in two places. One could make the argument that, since the code is the same, it should detect pulses the same and it is sufficient to check the range only on either P1 or P3. I agree this would be a valid argument to a certain extent, if the code had been written such that the P1 and P3 pulse-width checking were performed by the same code, not a copy, in two places, then I probably would not test both cases. However, since the code is copied, things can go wrong with copying the code in two places, such as forgetting to change a signal's name that applies to one pulse and not the other. This type of error can be hard to find, because the name is valid but not used in the correct place. And, of course, the code would check the correct section leaving this error to cause problems another day. So, I think it is a good idea to check both.

5.6. Stimulus

The input applied to the design is called the *stimulus*. It mimics the input data signals which are applied to the design by the simulator tool. Stimulus used in your simulation provides an easy way to observe the design's behavior within and outside your design's range without damaging the hardware. Stimulus can be provided by interactively typing it in real time, a graphical testbench, or an HDL testbench. There are advantages and disadvantages to each method. You can decide which is right for you.

5.6.1. Interactive Stimulus

Real-time input is typed on the command or transcript line of the simulator. The input data are not saved in a separate file. This means some information must be typed in between different simulation runs. All information is lost when the simulator is closed. This type of stimulus is not feasible for designs with a lot of inputs and, in my opinion, not feasible in most cases.

Fresh out of college, I was assigned to write a design in Advanced Boolean Equation Language (ABEL). I was not very familiar with either ABEL or programmable devices. The engineer taught me to test the code manually. This meant that, at least once a day, I had to retype all the input data. This was during the time when not many engineers had personal computers in their office, so I was working in a computer lab. This was a very time-consuming, painstaking exercise. In hindsight, I realize the engineer was not being mean but really did not know any better himself.

The `force` command is used in ModelSim to interactively set signal values, see the syntax shown in Example 5–1.

■ Example 5–1. Force Syntax

```
force [ -freeze | -drive | -deposit] [ -cancel <time>] [ -repeat <time>]
   <object_name> <value> [ <time>] [ , <value> <time> ...]
```
  where
- `freeze`
    Keeps the signal at a specific value until it is forced again or until it is unforced with a `noforce` command.
- `drive`
    A driver is attached to the signal and drives the specified value until the signal is forced again or until it is unforced with a `noforce` command.
- `deposit`
    Sets the signal to a specific value. This value stays the same until there is a subsequent driver transaction, until the signal is forced again, or until it is unforced with a `noforce` command.

(Continues)

> - <time>
>
>> Defines the time when the value is applied. The time is relative to the current simulation time unless an absolute time is stated by preceding the value with the @ character. The default resolution units are used if no time unit is specified. The change occurs in the current simulation delta cycle when the force command has a zero delay.
>>
>> ■

Use ModelSim's command line to type in the following four commands:

```
force clock20Mhz 0 25, 1 50 -repeat 50
```

```
noforce clock20Mhz
```

```
force reset 1 0, 0 125
```

```
noforce reset
```

```
force input_pulse 0 0, 1 175, 0 975, 1 5175, 0 5975
```

```
noforce input_pulse
```

```
run 7µsec
```

This creates a 20 MHz clock, sets and clears reset, creates P1 and P3 with 800 nsec pulse width, P1 to P3 pulse spacing of 5 µsec, and the simulation runs for 7 µsec., see Figure 5–4. Because the stimulus is not saved, these commands have to be entered each time the design is recompiled.

5.6.2. Graphical Test Bench

A graphical testbench uses waveforms to describe the behavior of the input signals. Like waveform designs, graphical testbenches are not as flexible and generally not portable to other simulation tools. In general, waveform editors are easier; however, you give up flexibility. Altera's Quartus II 8.1, Web Edition, is used to create the input stimulus shown in Figure 5–5, which represents the input signals

```
Clock20Mhz = 20MHz
```

```
Reset = 75 nsec
```

```
P1 pulsewidth = 850.00 nsec
```

```
P3 pulsewidth = 870.00 nsec
```

```
P1 to P3 pulse spacing = 5.2µsec
```

After applying the graphical stimulus to the design, P1 and P3 are detected and considered valid pulses. Since this satisfies the mode decode condition, a Mode 2 is decoded, see Figure 5–6.

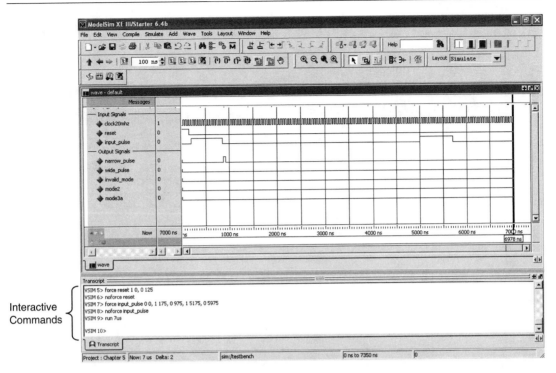

Figure 5-4: Interactive Stimulus Using ModelSim

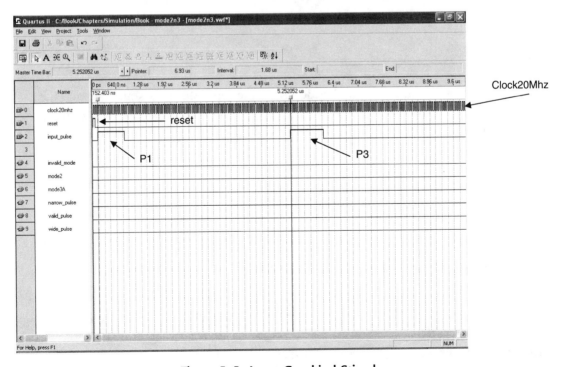

Figure 5-5: Input Graphical Stimulus

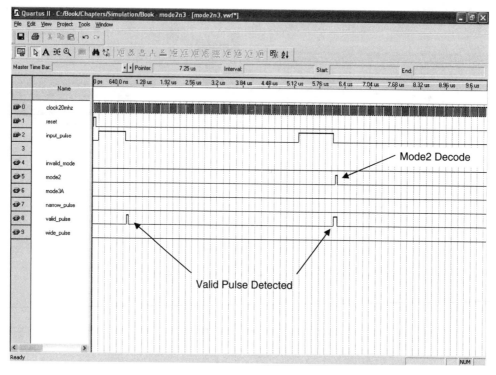

Figure 5–6: Output Results from Graphical Stimulus

5.6.3. HDL Testbench

An HDL testbench is an HDL file that describes the input. It looks similar to an HDL design and shares some of the same advantages. For example, it is easy to switch between different manufacturers and more flexible than graphical or interactive stimulus formats.

Testbenches can be written as

- **Manual**. The output results must be viewed manually to determine if they are correct.

- **Automatic**. Outputs are evaluated by the code and the final results are provided. Final results can be something like a pass/fail indicator on the screen or data written to an external file.

Each of these testbench options is examined in this section. A VHDL testbench has a design structure similar to the design code, it has the same sections as a regular VHDL design. A testbench starter template has been provided, see Listing 5–1.

Listing 5-1. VHDL Testbench Starter Template

```
1.    --*********************** Header Section ************************
2.    -- Name            : Rebecca C. Smith
3.    -- Date            : January 28, 2009
4.    -- Filename        : tb_EntityName.vhd
5.    --Description      : This starter HDL template provides placeholders and
      syntax that can be used
6.    --                 : to help develop VHDL testbenches. Modify the template
      to meet your needs.
7.    -- Revision History
8.    -- Date              Initials        Description
9.    -- ********************* End Header Section ***********************
10.   Library IEEE;                    -- define library and packages needed for
      this design
11.   Use IEEE.std_logic_1164.All;
12.
13.   Entity <entity name> Is End <entity name>;
14.
15.   Architecture <architecture name> Of <entity name> Is
16.   Component <component's name> Port (
17.   Signal <signal name>    : direction <data type>;    -- signal names in
   component's entity section
18.   End Component <entity name>;
19.
20.   Signal <signal's name>       : <data type>;
21.   Constant <constant's name>   : <data type>;
22.
23.   Begin
24.
25.   User's defined component name:  <Component's name>
26.   Port Map (
27.     Component1 signal name => user defined signal name,
28.     Component2 signal name => user defined signal name);
29.
30.   -- At this point you could have a combination of code to describe signal
      behaviors such as processes and signal assignments. This will be
      demonstrated by example.
31.   <process name>: Process (sensitivity list)       -- add process if
      necessary
32.
33.   Begin
34.     <sequential statements>;
35.   End Process;
36.   End <architecture name>;
```

Lines 1–9. Optional Heading

Lines 10–11. Library Section

The library section has the same meaning as in the design code.

Line 13. Entity Section

The entity section is a single line and has no signal names, as in the design code. As with the design code, it is a good idea to develop a naming convention for your testbenches. I discovered that naming all my testbench entities *testbench* made it easy for me to quickly locate the testbench in my simulation tool. This came about because once my design had over 30 files with at least 10 different testbenches. It was a nightmare trying to locate the correct testbench because of the way the simulator listed the files. So I learned that, if I name all testbench entities *testbench* it is easy for me to find a specific testbench. All I have to do is select the entity named testbench then the specific architecture. This makes my life a lot easier.

Lines 15–36. Architecture Section

The architecture can be a little confusing. It is the same general concept as the design phase's architecture section, which describes the design; however, this one defines the input signals or stimulus. My architecture's name is defined such that it gives an indication to what it is verifying.

Consider the testbench as a breadboard with a socket being used to test a chip, see Figure 5–7. Here is a simple scenario for testing the chip excluding power suppliers: Insert the chip into the socket, set up the data generator to provide input data, connect the data generator to the circuit using probes, and connect a logic analyzer to view the circuit's response. This is very similar to what is going on in the architecture. In the architecture section, the design can be thought of as a `component`; it is instantiated (put in the socket) in the testbench design; the input stimulus and testbench design are connected using internally defined signals; and the output is viewed using the simulation tool.

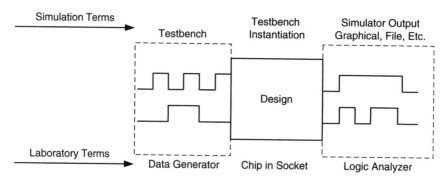

Figure 5–7: Simulation Phase Terms Equated to Lab Terms

Much like our lab setup, we can test different chip designs using the same setup for equivalent pinouts, just by removing one and inserting another. While the same is true for the testbench, if you have another variation of the design, the same testbench can be used just by replacing which design is instantiated.

Running the testbench is like turning on the data generator, because the inputs are then applied to the design and can be observed as waveforms, data listing (i.e., binary, hex, etc.), or results written to a file or your monitor.

5.6.4. Manual Testbench

The results from a manual testbench must be manually reviewed to determine if they are correct.

Design verifications performed this way can be slow; however, it is my option that beginners learn more from manual testbenches. While painstaking as this can be, I always found it exciting because I can explore the various features of my simulator and observe internal signal functions or other aspects of the design that may otherwise be overlooked with an automatic testbench. By using manual testbenches, I have found many design errors by being able to view internal signals that may have been overlooked. Personally, I like looking at the waveforms and internal signals, especially if I am working with someone else's code. The testbench shown in Listing 5–2 is used to verify that the design meets the requirements specified in the design package. Notice that the testbench has the same four sections as the design, with the information in the entity and architecture being slightly different from the design.

Lines 1–17. Optional Heading Section

The same general information is contained in this section, with the description describing the test stimulus.

Lines 19–20. Library Declaration Section

Only the `std_logic_1164` package from the IEEE library is needed for this testbench, so it is made visible and usable in the library section.

Line 22. Entity Section

For the testbench, input and output signals are not called out in this section. Only this single line is required.

Lines 24–173. Architecture Section

Lines 26–35. Component Declaration

The `mode2n3` design is defined as a component.

Listing 5-2. Manual Testbench

```
1.    -- ******************** Header Section ****************************
2.    -- Name                  : James W. Smith
3.    -- Date                  : June 4, 2009
4.    -- Filename              : tb_mode2n3.vhd
5.    -- Description           : This testbench determines if the input pulses
      meets the Mode 2 or 3A
6.    --                       : pulse width and spacing requirements by:
7.    --                       : 1. verifying pulse widths are .7µsec to .9µsec
      inclusive
8.    --                       : 2. Mode 2 P1 to P3 pulse spacing is 4.9µsec to
      5.1µsec inclusive
9.    --                       : 3. Mode 3A P1 to P3 pulse spacing is 7.9µsec to
      7.1µsec inclusive
10.   --                       :
11.   --                       : Results must be manually verified.
12.   --                       : Run simulation for 500.00µsec
13.   --
14.   -- Revision History
15.   -- Date                  Initials            Description
16.   --
17.   --**************** End Header Section ********************
18.
19.   Library  IEEE;                        -- define  library  and  packages
      needed for this design
20.   Use IEEE.std_logic_1164.All;
21.
22.   Entity testbench Is End testbench;
23.
24.   Architecture tb_mode2n3 Of testbench Is
25.
26.   Component mode2n3 Port (
27.      clock20Mhz            : In std_logic;
28.      reset                : In std_logic;
29.      input_pulse          : In std_logic;
30.      narrow_pulse         : Out std_logic;
31.      wide_pulse           : Out std_logic;
32.      invalid_mode         : Out std_logic;
33.      mode2                : Out std_logic;
34.      mode3A               : Out std_logic);
35.   End Component mode2n3;
36.
37.   Signal clock20Mhz         : std_logic := '0' ;
38.   Signal reset              : std_logic := '1' ;
39.   Signal input_pulse        : std_logic := '0' ;
```

```
40.   Signal narrow_pulse              : std_logic;
41.   Signal wide_pulse               : std_logic;
42.   Signal invalid_mode             : std_logic;
43.   Signal mode2                    : std_logic;
44.   Signal mode3A                   : std_logic;
45.
46.   Constant twenty_five_nsec        : time := 25 nsec;
47.
48.   Begin
49.
50.   mode2n3_component: mode2n3
51.   Port Map (
52.      clock20Mhz       => clock20Mhz,
53.      reset            => reset,
54.      input_pulse      => input_pulse,
55.      narrow_pulse     => narrow_pulse,
56.      wide_pulse       => wide_pulse,
57.      invalid_mode     => invalid_mode,
58.      mode2            => mode2,
59.      mode3A           => mode3A);
60.
61.   create_twenty_Mhz: Process
62.   Begin
63.      Wait For twenty_five_nsec;
64.         clock20Mhz     <= Not clock20Mhz;
65.   End Process;
66.
67.   reset                <=   '0' After 145.00 nsec;
68.
69.   input_pulse          <=
70.             -- Test Case 1 P1 pulse width (pw) = narrow & no P3
71.             '1' After 200.00 nsec,       -- P1, Test Pulse 1
72.             '0' After 850.00 nsec,       -- 650nsec narrow pulse width
73.
74.             -- Test Case 2 P1 PW = min.; P3 PW = narrow & M2 normal spacing
75.             '1' After 5.00 µsec,         -- P1, Test Pulse 2
76.             '0' After 5.70 µsec,         -- 700nsec min pulse width
77.
78.             '1' After 10.00 µsec,        -- P3, Test Pulse 3
79.             '0' After 10.65 µsec,        -- 650nsec pulse width
80.
81.             -- Test Case 3 P1 PW normal; P3 PW min & M2 normal spacing
82.             '1' After 35.00 µsec,        -- P1, Test Pulse 4
83.             '0' After 35.80 µsec,        -- 800nsec max pulse width
84.
```

```
85.            '1' After 40.00 µsec,     -- P3, Test Pulse 5
86.            '0' After 40.70 µsec,     -- 700nsec wide pulse width
87.

88.            -- Test Case 4 P1 PW = max, P3 PW = normal & M2 min spacing
89.            '1' After 60.00 µsec,     -- P1, Test Pulse 6
90.            '0' After 60.90 µsec,     -- 900nsec pulse width
91.

92.            '1' After 64.90 µsec,     -- P3, Test Pulse 7
93.            '0' After 65.70 µsec,     -- 800nsec pulse width
94.

95.            -- Test Case 5 P1 PW = wide; no P3
96.            '1' After 90.00 µsec,     -- P1, Test Pulse 8
97.            '0' After 90.95 µsec,     -- 950nsec pulse width
98.

99.            -- Test Case 6 P1 PW = normal; P3 PW = max & M2 max spacing
100.           -- Mode 2 5.1µsec P1 - P3 spacing
101.           '1' After 110.10 µsec,      -- P1, Test Pulse 9
102.           '0' After 110.90 µsec,      -- 800nsec pulse width
103.

104.           '1' After 115.20 µsec,      -- P3, Test Pulse 10
105.           '0' After 116.10 µsec,      -- 900nsec pulse width
106.

107.           -- Test Case 7 P1 PW = normal & P3 PW = wide & M2 normal spacing
108.           '1' After 136.00 µsec,      -- P1, Test Pulse 11
109.           '0' After 136.80 µsec,      -- 800nsec pulse width
110.

111.           '1' After 141.0 µsec,       -- P3, Test Pulse 12
112.           '0' After 141.95 µsec,      -- 950nsec pulse width
113.

114.           -- Test Case 8 P1 PW = normal; P3 PW = normal & M2 narrow spacing
115.           '1' After 160.00 µsec,      -- P1, Test Pulse 13
116.           '0' After 160.80 µsec,      -- 800nsec pulse width
117.

118.           '1' After 164.80 µsec,      -- P3, Test Pulse 14
119.           '0' After 165.60 µsec,      -- 800nsec pulse width
120.

121.           -- Test Case 9 P1 PW = normal; P3 PW = normal & M2 wide spacing
122.           '1' After 180.00 µsec,      -- P1, Test Pulse 15
123.           '0' After 180.80 µsec,      -- 800nsec pulse width
124.

125.           '1' After 185.20 µsec,      -- P3, Test Pulse 16
126.           '0' After 186.00 µsec,      -- 800nsec pulse width
127.
128.  _____
129.

130.           -- Test Case 10 P1 PW = normal; P3 PW = normal & M3 narrow spacing
```

```
131.                '1' After 300.00 µsec,        -- P1, Test Pulse 17
132.                '0' After 300.80 µsec,        -- 800nsec pulse width
133.
134.                '1' After 307.80 µsec,        -- P3, Test Pulse 18
135.                '0' After 308.60 µsec,        -- 800nsec pulse width
136.
137.                -- Test Case 11 P1 PW = normal; P3 PW = normal & M3 min spacing
138.                '1' After 320.00 µsec,        -- P1, Test Pulse 19
139.                '0' After 320.80 µsec,        -- 800nsec pulse width
140.
141.                '1' After 327.90 µsec,        -- P3, Test Pulse 20
142.                '0' After 328.70 µsec,        -- 800nsec pulse width
143.
144.               -- Test Case 12 P1 PW = normal; P3 PW = normal & M3 normal spacing
145.                '1' After 340.00 µsec,        -- P1, Test Pulse 21
146.                '0' After 340.80 µsec,        -- 800nsec pulse width
147.
148.                '1' After 348.00 µsec,        -- P3, Test Pulse 22
149.                '0' After 348.80 µsec,        -- 800nsec pulse width
150.
151.                -- Test Case 13 P1 PW = normal; P3 PW = normal & M3 max spacing
152.                '1' After 360.00 µsec,        -- P1, Test Pulse 23
153.                '0' After 360.80 µsec,        -- 800nsec pulse width
154.
155.                '1' After 368.10 µsec,        -- P3, Test Pulse 24
156.                '0' After 368.90 µsec,        -- 800nsec pulse width
157.
158.                -- Test Case 14 P1 PW = normal; P3 PW = normal & M3 wide spacing
159.                '1' After 380.00 µsec,        -- P1, Test Pulse 25
160.                '0' After 380.80 µsec,        -- 800nsec pulse width
161.
162.                '1' After 388.20 µsec,        -- P3, Test Pulse 26
163.                '0' After 389.00 µsec,        -- testing at 8.2µsec spacing
164.
165.   -- Out of range area between Modes 2 & 3 pulse spacing 5.8µsec & 7.8µsec;
      pulse spacing = 6µsec
166.
167.                '1' After 480.00 µsec,        -- normal P1, Test Pulse 27
168.                '0' After 480.80 µsec,
169.
170.                '1' After 486.00 µsec,        -- normal P3, Test Pulse 28
171.                '0' After 486.80 µsec;
172.
173.   End tb_mode2n3;
```

Note: Lines 37–171 are referring to Listing 5–2.

Lines 37–44. Defining Internal Signals

All the internal signals used in this testbench are defined in this section. For the power-on `reset` signal on the board it is used to set inputs to a known state, the initial conditions must be set for simulation. The initial conditions for the input signals are indicated by appending := 'X' ; , where X is any valid signal level (i.e., 1, 0, Z, etc.) to the signal definition. The initial condition for `clock20Mhz` and `input_pulse` is low or '0' and `reset` is set high or '1'.

Line 46. Defining Half Clock Period

This constant is used to create the 20 MHz clock.

Lines 50–59. Instantiating the Design

These lines instantiate the 'mode2n3' design. Internal signals are used to connect the design's IO. To keep things simple, I like to make my connecting signal's name the same as those on the component. However, it is acceptable to make them different.

Lines 61–65. Creating a 20 MHz Clock

There are several ways to create a repetitive signal like the clock. I like to use a process that toggles the signal every half cycle.

Line 38. Reset Signal

This is the power-on `reset` signal, which is initially set active or high. Keeping `reset` active, the outputs do not respond to changes in the input, see Figure 5–8. To create this scenario, line 67, which sets `reset` inactive, was commented out.

Line 67. Inactive `reset`

This line, read as `reset`, goes low after 145 nsec. Now the outputs respond to the input stimulus, see Figure 5–9. `Reset` is active only once in this design; however, it could be reactivated at any time during the simulation.

Lines 69–171. `Input_pulse` *signal*

The P1 and P3 are received through `input_pulse` signal. The test cases start on line 71. Using the predefined test cases, the various pulse widths and spacings are described.

5.6.5. Simulation Phase Outputs

The output is shown as a waveform for this testbench. The full simulation waveform that results from applying the input stimulus defined in the testbench is shown in Figure 5–10. The full simulation view is difficult to read, so the expanded view of test case 1, test case 4, and test case 14 are presented in this section.

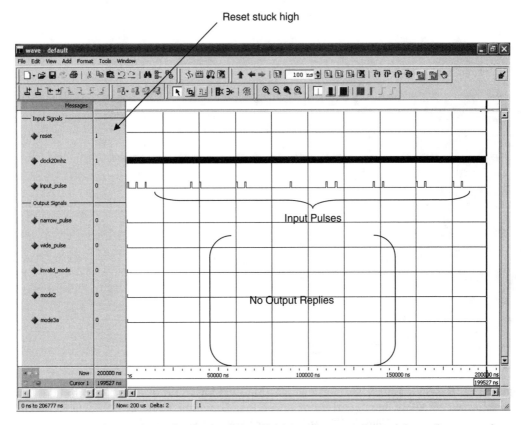

Figure 5–8: Continuously Active Reset (Material based on or adapted from figures and text owned by Xilinx, Inc., courtesy of Xilinx, Inc. Copyright Xilinx © 1995–2008 used in Xilinx ISE WebPack™ software version 10.1.)

Lines 70–72. Test Case 1

P1 pulse goes high at 200.00 nsec, then low at 850.00 nsec. This produces a 650 nsec pulse. The minimum pulse width for a mode decode is 0.7 sec; therefore, a narrow pulse width goes active on the falling edge of P1 and no P3 is sent, see Figure 5–11.

Lines 88–93. Test Case 4

P1 is a 900 sec pulse width, rising edge at 60.00 sec and falling edge at 60.9 sec. P3 is 800 nsec pulse width, rising edge at 64.9 sec and falling edge at 65.70 sec. The rising edge timing between P1 and P3 is 4.9 sec, the minimum spacing for Mode 2, see Figure 5–12.

Lines 158–163. Test Case 14

P1 and P3 have normal pulse widths and the rising edge timing between P1 and P3 is 8.2 sec, which is wide or outside the maximum pulse spacing for Mode 3A, see Figure 5–13.

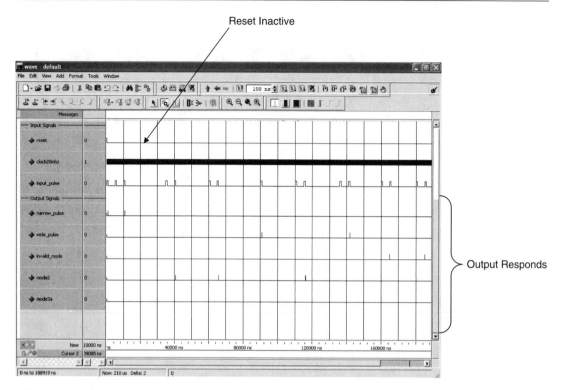

Figure 5–9: Reset Activated and Deactivated (Material based on or adapted from figures and text owned by Xilinx, Inc., courtesy of Xilinx, Inc. Copyright Xilinx © 1995–2008 used in Xilinx ISE WebPack™ software version 10.1.)

5.6.6. Automatic Testbench

An automatic testbench can make design verification easier. However, if your automatic testbench is not complete or correct, the results can be misleading, in that you may be searching for errors in the design when in reality they are in the testbench. With an automatic testbench, you are not required to manually view or verify the output. This type of testbench is designed to evaluate the output results and provide the final results.

However, just like anything else, you have to make sure that this testbench is reporting accurate information. Otherwise, the results can be wrong. The manual testbench has been modified to write the test results to an external file. The output file should contain whatever information is necessary to verify that the design meets the requirements. For the automatic testbench shown in Listing 5–3, I decided I wanted to see if the design was able to detect wide, narrow, or valid pulse widths and P1 or P3 pulse spacing. Someone else may decide it is acceptable to write only when a valid mode is decoded and assume the other input pulses did not create a mode decode. Using the `read` and `write` commands in testbenches can make verification a lot easier.

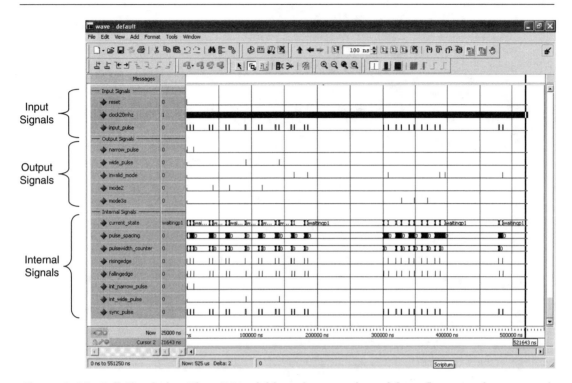

Figure 5-10: Full Simulation View (Material based on or adapted from figures and text owned by Xilinx, Inc., courtesy of Xilinx, Inc. Copyright Xilinx © 1995–2008 used in Xilinx ISE WebPack™ software version 10.1.)

Changes are made to the manual testbench to make it automatic.

Lines 1–21. Optional Heading Section

The description is updated to include the name of the file where the simulation results will be written.

Lines 23–26. Library Declaration

The `write` command used to write to the output file, which is located in the std.textio package in the std library. Therefore, this library is called out and made visible to the code.

Line 28. Entity

This is the same as in the manual testbench.

Lines 30–246. Architecture Section

To distinguish between the manual and automatic architecture, the name `auto` is appended to the end of the architecture's name. This section is basically the same as the manual

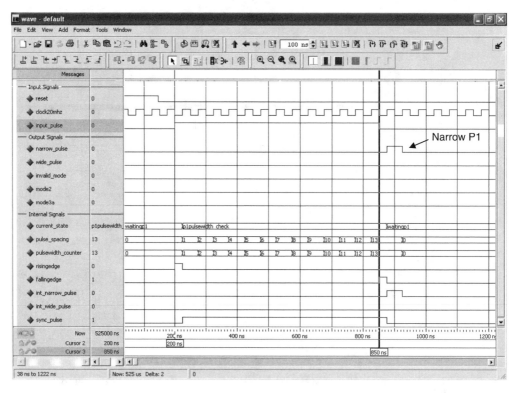

Figure 5–11: Test Case 1, Simulated Output (Material based on or adapted from figures and text owned by Xilinx, Inc., courtesy of Xilinx, Inc. Copyright © Xilinx 1995–2008 used in Xilinx ISE WebPack™ software version 10.1.)

testbench, with the addition of defining the file where the output will be written and the condition in which data are written to the file.

Line 58. Defining the Output File

This defines the file `results_file.txt`, where the output results will be written.

Lines 89–138. Writing the Output Data Process

This process uses the output signals `narrow_pulse`, `wide_pulse`, `invalid_mode`, `mode2`, `mode3A`, and `valid_pulse` to determine the status of the input signal. A message indicating the status is written to the output file. For example, if the input signal is narrow or outside the required pulse width range, a message similar to `Test pulse X is narrow` is written to the output file. X represents the number of the test pulse. For each test case, there is a comment stating the test pulse number for P1 or P3. By examining the output file, I can determine if the design is able to detect narrow, wide, or valid pulse widths and pulse spacings.

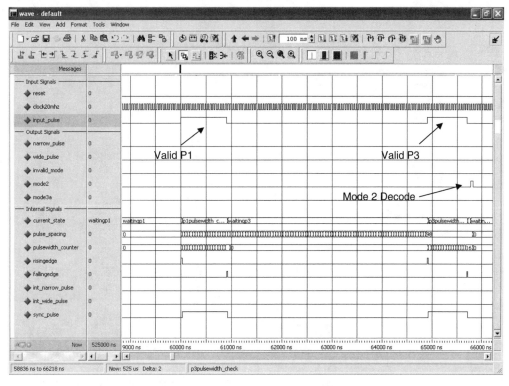

Figure 5–12: Test Case 4, Simulated Output (Material based on or adapted from figures and text owned by Xilinx, Inc., courtesy of Xilinx, Inc. Copyright © Xilinx 1995–2008 used in Xilinx ISE WebPack™ software version 10.1.)

Automatic Testbench Output Results

After applying the stimulus to the automatic testbench, the output is written in `results_file.txt`, see Example 5–2. This file is written in the same directory as my simulation work directory created by the simulator.

5.6.7. Capture Data

At times, trying to troubleshoot a fielded design problem or manually create a test stimulus is not feasible. If you find yourself in this situation, then utilizing test equipment can be a good option. Tektronix offers a series of data capture and acquisition equipment with offline viewers that can be used on your PC. Using Tektronix's logic analyzer, real-time data can be captured, stored, and exported as a text file. The captured data can be read into a testbench for simulation and imported, viewed, and modified (if necessary) using TLA® Application, the PC interface offline viewer. Setups created using TLA Application can be exported to a logic analyzer.

Similar data patterns created using PG3A Series Digital Pattern Generator can be exported, viewed, and modified (if necessary) using the offline PC viewer PGAppDotNet. Such data

■ Example 5–2. Output Results

```
Test pulse 1 is narrow
Test pulse 2 is valid
Test pulse 3 is narrow
Test pulse 4 is valid
Test pulse 5 is valid
P1 or test pulse 4 to P3 or test pulse 5 spacing is a Mode2
Test pulse 6 is valid
Test pulse 7 is valid
P1 or test pulse 6 to P3 or test pulse 7 spacing is a Mode2
Test pulse 8 is wide
Test pulse 9 is valid
Test pulse 10 is valid
P1 or test pulse 9 to P3 or test pulse 10 spacing is a Mode2
Test pulse 11 is valid
Test pulse 12 is wide
Test pulse 13 is valid
P1 or test pulse 13 to P3 or test pulse 14 spacing is an invalid mode
Test pulse 15 is valid
P1 or test pulse 15 to P3 or test pulse 16 spacing is an invalid mode
Test pulse 17 is valid
P1 or test pulse 17 to P3 or test pulse 18 spacing is an invalid mode
Test pulse 19 is valid
Test pulse 20 is valid
P1 or test pulse 19 to P3 or test pulse 20 spacing is a Mode3A
Test pulse 21 is valid
Test pulse 22 is valid
P1 or test pulse 21 to P3 or test pulse 22 spacing is a Mode3A
Test pulse 23 is valid
Test pulse 24 is valid
P1 or test pulse 23 to P3 or test pulse 24 spacing is a Mode3A
Test pulse 25 is valid
P1 or test pulse 25 to P3 or test pulse 26 spacing is an invalid mode
Test pulse 27 is valid
P1 or test pulse 27 to P3 or test pulse 28 spacing is an invalid mode                ■
```

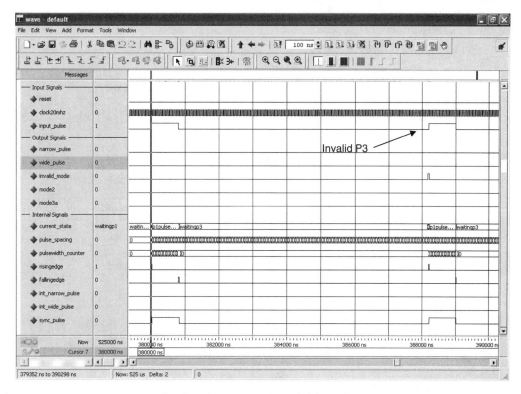

Figure 5–13: Test Case 14, Simulated Output (Material based on or adapted from figures and text owned by Xilinx, Inc., courtesy of Xilinx, Inc. Copyright © Xilinx 1995–2008 used in Xilinx ISE WebPack™ software version 10.1.)

can be read into a testbench for simulation. Setups created using PGAppDotNet can be exported to the pattern generator.

The offline viewers are great options, especially when you share lab equipment. By using the offline viewers, you can reduce bench time by capturing data and working with it at your desk. I really like working in the lab and at times prefer it over being at my desk. However, many times, I had to utilize the offline interface due to sharing equipment. Not having total access to capture data meant I had to make the most of my resource, which included the PC offline interface. I was able to set up or make changes to my equipment setup and utilize saved data at my desk. My lab time was spent capturing data I could utilize offline. The offline viewers give you the same look and feel as being in the lab, but you are actually at your desk.

An example of a waveform using TLA Application is shown in Figure 5–14. Signals on the waveform can be shown as binary, hex, octal, decimal, singed decimal, or symbolic. They can be viewed in groups or as individual signals. Many features and options are available when using TLA Application. I suggest downloading the free PC interface by going to www.tek.com and searching for TLA Application.

Listing 5–3. Automatic Testbench

```
1.   --*************************** Header Section ***************************
2.   -- Name              : James W. Smith
3.   -- Date              : August 25, 2009
4.   -- Filename          : tb_mode2n3_auto.vhd
5.   -- Description       : This testbench determines if the input pulses meets the
     Mode 2 or 3A
6.   --                   : pulse width and spacing requirements by:
7.   --                   : 1. Verifying pulse widths are .7µsec to .9µsec inclusive
8.   --                   : 2. Mode 2 P1 to P3 pulse spacing is 4.9µsec to 5.1µsec
     inclusive
9.   --                   : 3. Mode 3A P1 to P3 pulse spacing is 7.9µsec to 7.1µsec
     inclusive
10.  --                   :
11.  --                   : Simulation results will be written to "results_file.
     txt".
12.  --                   : Results file indicates if:
13.  --                   : Input pulse width is narrow, wide or valid
14.  --                   : P1 to P3 pulse spacing for M2 or M3A is valid or invalid
15.  --
16.  --                   : Run simulation for 500.00µsec
17.  --
18.  -- Revision History
19.  -- Date            Initials        Description
20.  --
21.  --*************************** End Header Section ********************
22.  --
23.  Library IEEE;              -- define library and packages needed for this
     design
24.  Use IEEE.std_logic_1164.All;
25.  Library Std;
26.  Use std.textio.All;
27.
28.  Entity testbench Is End testbench;
29.
30.  Architecture tb_mode2n3_auto Of testbench Is
31.
32.  Component mode2n3 Port (
33.    clock20Mhz              : In std_logic;
34.    reset                   : In std_logic;
35.    input_pulse             : In std_logic;
36.    narrow_pulse            : Out std_logic;
37.    wide_pulse              : Out std_logic;
38.    invalid_mode            : Out std_logic;
39.    valid_pulse             : Out std_logic;
```

```
40.    mode2                 : Out std_logic;
41.    mode3A                : Out std_logic);
42. End Component mode2n3;
43.
44. Signal clock20Mhz        : std_logic := '0';
45. Signal reset             : std_logic := '1';
46. Signal input_pulse       : std_logic := '0';
47. Signal narrow_pulse      : std_logic;
48. Signal wide_pulse        : std_logic;
49. Signal invalid_mode      : std_logic;
50. Signal valid_pulse       : std_logic;
51. Signal mode2             : std_logic;
52. Signal mode3A            : std_logic;
53.
54. Signal pulse_number      : integer;
55.
56. Constant twenty_five_nsec  : time := 25 nsec;
57.
58. File data_out: Text Open write_mode Is "results_file.txt";
59.
60. Begin
61.
62. mode2n3_component: mode2n3
63. Port Map(
64.    clock20Mhz            => clock20Mhz,
65.    reset                 => reset,
66.    input_pulse           => input_pulse,
67.    narrow_pulse          => narrow_pulse,
68.    wide_pulse            => wide_pulse,
69.    invalid_mode          => invalid_mode,
70.    valid_pulse           => valid_pulse,
71.    mode2                 => mode2,
72.    mode3A                => mode3A);
73.
74. create_twenty_Mhz: Process
75. Begin
76.    Wait For twenty_five_nsec;
77.       clock20Mhz   <= Not clock20Mhz;
78. End Process;
79.
80. count_test: Process (reset, input_pulse)
81. Begin
82.    If reset = '1' Then
83.       pulse_number  <= 0;
84.    Elsif rising_edge (input_pulse) Then
85.       pulse_number  <= pulse_number + 1;
```

```vhdl
86.    End If;
87. End Process;
88.
89. write_results: Process (clock20Mhz)
90. Variable data_line  : line;
91.
92. Begin
93.    If rising_edge(clock20Mhz) Then
94.      If narrow_pulse = '1' Then
95.         writeline (data_out, data_line);
96.         write (data_line, string'("Test pulse "));
97.         write (data_line, pulse_number);
98.         write (data_line, string'(" is narrow"));
99.         writeline (data_out, data_line);
100.      Elsif wide_pulse = '1' Then
101.         writeline (data_out, data_line);
102.         write (data_line, string'("Test pulse "));
103.         write (data_line, pulse_number);
104.         write (data_line, string'(" is wide"));
105.         writeline (data_out, data_line);
106.      Elsif invalid_mode = '1' Then
107.         writeline (data_out, data_line);
108.         write (data_line, string'("P1 or test pulse"));
109.         write (data_line, pulse_number - 1);
110.         write (data_line, string'("to P3 or test pulse"));
111.         write (data_line, pulse_number);
112.         write (data_line, string'("spacing is an invalid mode"));
113.         writeline (data_out, data_line);
114.      Elsif mode2 = '1' Then
115.         writeline (data_out, data_line);
116.         write (data_line, string'("P1 or test pulse"));
117.         write (data_line, pulse_number - 1);
118.         write (data_line, string'(" to P3 or test pulse "));
119.         write (data_line, pulse_number);
120.         write (data_line, string'("spacing is a Mode2"));
121.         writeline (data_out, data_line);
122.      Elsif mode3A = '1' Then
123.         writeline (data_out, data_line);
124.         write (data_line, string'("P1 or test pulse "));
125.         write (data_line, pulse_number - 1);
126.         write (data_line, string'("to P3 or test pulse"));
127.         write (data_line, pulse_number);
128.         write (data_line, string'("spacing is a Mode3A"));
129.         writeline (data_out, data_line);
130.      Elsif valid_pulse = '1' Then
131.         writeline (data_out, data_line);
```

```
132.            write (data_line, string'("Test pulse "));
133.            write (data_line, pulse_number);
134.            write (data_line, string'(" is valid"));
135.            writeline (data_out, data_line);
136.        End If;
137.      End If;
138.  End Process;
139.
140.  reset            <=   '0' After 145.00 nsec;
141.
142.  input_pulse <=
143.                   -- Test Case 1 P1 pulse width (pw) = narrow & no P3
144.                   '1' After 200.00 nsec,    -- P1, Test Pulse 1
145.                   '0' After 850.00 nsec,    -- 650nsec narrow pulse width
146.
147.                   -- Test Case 2 P1 PW = min.; P3 PW = narrow & M2 normal spacing
148.                   '1' After 5.00 µsec,      -- P1, Test Pulse 2
149.                   '0' After 5.70 µsec,      -- 700nsec min pulse width
150.
151.                   '1' After 10.00 µsec,     -- P3, Test Pulse 3
152.                   '0' After 10.65 µsec,     -- 650nsec pulse width
153.
154.                   -- Test Case 3 P1 PW normal; P3 PW min & M2 normal spacing
155.                   '1' After 35.00 µsec,     -- P1, Test Pulse 4
156.                   '0' After 35.80 µsec,     -- 800nsec max pulse width
157.
158.                   '1' After 40.00 µsec,     -- P3, Test Pulse 5
159.                   '0' After 40.70 µsec,     -- 700nsec wide pulse width
160.
161.                   -- Test Case 4 P1 PW = max, P3 PW = normal & M2 min spacing
162.                   '1' After 60.00 µsec,     -- P1, Test Pulse 6
163.                   '0' After 60.90 µsec,     -- 900nsec pulse width
164.
165.                   '1' After 64.90 µsec,     -- P3, Test Pulse 7
166.                   '0' After 65.70 µsec,     -- 800nsec pulse width
167.
168.                   -- Test Case 5 P1 PW = wide; no P3
169.                   '1' After 90.00 µsec,     -- P1, Test Pulse 8
170.                   '0' After 90.95 µsec,     -- 950nsec pulse width
171.
172.                   -- Test Case 6 P1 PW = normal; P3 PW = max & M2 max spacing
173.                   -- Mode 2 5.1µsec P1 - P3 spacing
174.                   '1' After 110.10 µsec,    -- P1, Test Pulse 9
175.                   '0' After 110.90 µsec,    -- 800nsec pulse with
176.
177.                   '1' After 115.20 µsec,    -- P3, Test Pulse 10
```

```
178.          '0' After 116.10 µsec,      -- 900nsec pulse width
179.
180.          -- Test Case 7 P1 PW = normal & P3 PW = wide & M2 normal spacing
181.          '1' After 136.00 µsec,      -- P1, Test Pulse 11
182.          '0' After 136.80 µsec,      -- 800nsec pulse width
183.
184.          '1' After 141.0 µsec,       -- P3, Test Pulse 12
185.          '0' After 141.95 µsec,      -- 950µsec pulse width
186.
187.          -- Test Case 8 P1 PW = normal; P3 PW = normal & M2 narrow spacing
188.          '1' After 160.00 µsec,      -- P1, Test Pulse 13
189.          '0' After 160.80 µsec,      -- 800nsec pulse width
190.
191.          '1' After 164.80 µsec,      -- P3, Test Pulse 14
192.          '0' After 165.60 µsec,      -- 800µsec pulse width
193.
194.          -- Test Case 9 P1 PW = normal; P3 PW = normal & M2 wide spacing
195.          '1' After 180.00 µsec,      -- P1, Test Pulse 15
196.          '0' After 180.80 µsec,      -- 800nsec pulse width
197.
198.          '1' After 185.20 µsec,      -- P3, Test Pulse 16
199.          '0' After 186.00 µsec,      -- 800µsec pulse width
200.
201.   -- -- -- -- -- -- -- -- -- -- -- -- -- -- -- -- --
202.
203.          -- Test Case 10 P1 PW = normal; P3 PW = normal & M3 narrow spacing
204.          '1' After 300.00 µsec,      -- P1, Test Pulse 17
205.          '0' After 300.80 µsec,      -- 800nsec pulse width
206.
207.          '1' After 307.80 µsec,      -- P3, Test Pulse 18
208.          '0' After 308.60 µsec,      -- 800µsec pulse width
209.
210.          -- Test Case 11 P1 PW = normal; P3 PW = normal & M3 min spacing
211.          '1' After 320.00 µsec,      -- P1, Test Pulse 19
212.          '0' After 320.80 µsec,      -- 800nsec pulse width
213.
214.          '1' After 327.90 µsec,      -- P3, Test Pulse 20
215.          '0' After 328.70 µsec,      -- 800µsec pulse width
216.
217.          -- Test Case 12 P1 PW = normal; P3 PW = normal & M3 normal spacing
218.          '1' After 340.00 µsec,      -- P1, Test Pulse 21
219.          '0' After 340.80 µsec,      -- 800nsec pulse width
220.
221.          '1' After 348.00 µsec,      -- P3, Test Pulse 22
222.          '0' After 348.80 µsec,      -- 800µsec pulse width
223.
```

```
224.                    -- Test Case 13 P1 PW = normal; P3 PW = normal & M3 max spacing
225.            '1' After 360.00 µsec,      -- P1, Test Pulse 23
226.            '0' After 360.80 µsec,      -- 800nsec pulse width
227.
228.            '1' After 368.10 µsec,      -- P3, Test Pulse 24
229.            '0' After 368.90 µsec,      -- 800µsec pulse width
230.
231.                    -- Test Case 14 P1 PW = normal; P3 PW = normal & M3 wide spacing
232.            '1' After 380.00 µsec,      -- P1, Test Pulse 25
233.            '0' After 380.80 µsec,      -- 800nsec pulse width
234.
235.            '1' After 388.20 µsec,      -- P3, Test Pulse 26
236.            '0' After 389.00 µsec,      -- testing at 8.2µsec spacing
237.
238.    -- Out of range area between Modes 2 & 3 pulse spacing 5.8µsec & 7.8µsec;
    pulse spacing = 6µ
239.
240.            '1' After 480.00 µsec,      -- normal P1, Test Pulse 27
241.            '0' After 480.80 µsec,
242.
243.            '1' After 486.00 µsec,      -- normal P3, Test Pulse 28
244.            '0' After 486.80 µsec;
245.
246.    End tb_mode2n3_auto;
```

If your simulator can read the waveform output from the analyzer, you can directly import it into your simulation. However, my simulator cannot read the waveform output format, which is not a problem, because the data easily can be viewed as either a waveform or listing. So selecting the Listing option icon on the offline viewer, the waveform is now shown as data points, see Figure 5–15. This file can be saved as text and read into my testbench.

Some or all the data shown in the listing can be exported into a text file. The offline interface allows you to customize the exported text file. Some of the export data options are

- Space, tab, comma, or semicolon field delimiter.

- Enhanced column headers.

- Including or omitting column heading information.

- Including unit characters.

- Radix.

The 8 bits of A3 is exported to a text file. Because the text file is very large, only a very small portion is shown in Example 5–3. I removed the heading information and unit characters,

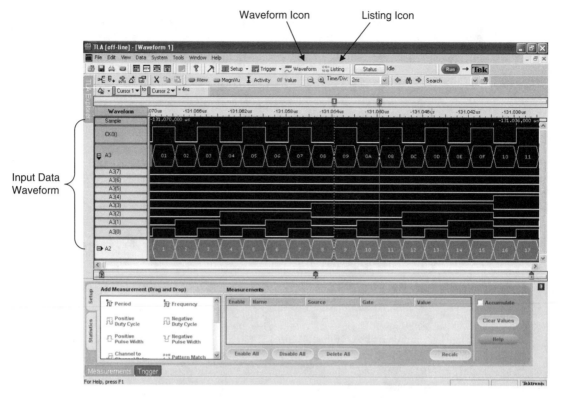

Figure 5-14: TLA Application Waveform (Screen shot taken from TLA Application Software V5.1 SP1 Offline Viewer courtesy of Tektronix, Inc.)

■ Example 5-3. Exported Listing Data

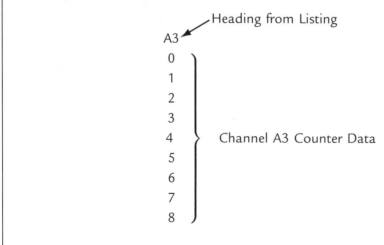

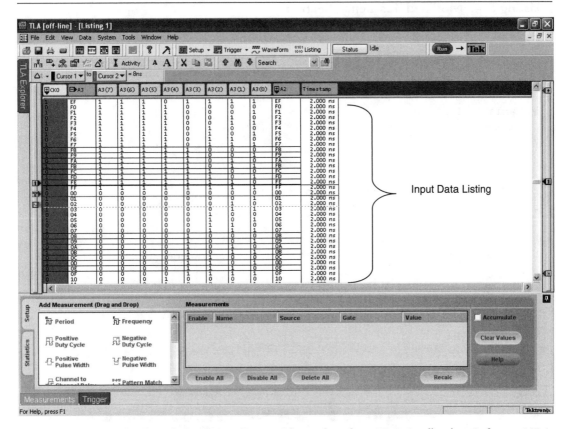

Figure 5–15: Application Listing Data (Screen shot taken from TLA Application Software V5.1 SP1 Offline Viewer courtesy of Tektronix, Inc.)

which make reading the file easier, but left the heading A3 for illustration purposes. The data were exported as decimal, because this is an acceptable input for the read command; however, it will be converted to `std_logic_vector` in the testbench. Channel A3 is a counter; this data will be read into the testbench and used as the stimulus for the clock, reset, and input pulse signal. This approach was selected because it demonstrates some additional things you can do when reading in data.

The testbench has been modified to read these data, but only the command that defines the file and the process that reads the file are shown, see Listing 5–4.

Using the text file data, the testbench gives the results shown in Figure 5–16.

Line 1. Defining `Read File`

The external file that contains the input data is defined as `TLA_Data.txt`. This command is inserted in the architecture section prior to the initial `Begin` statement.

Listing 5–4. Modified Testbench Section

```
1. File data_in: Text Open read_mode Is "TLA_Data.txt";        -- defines the
   file to be read
2.
3. read_file: Process
4.
5. Variable data_line    : line;
6. Variable data_integer : integer;
7.
8. Begin
9.
10.  While Not endfile(data_in) Loop
11.     readline (data_in, data_line);
12.
13.     read (data_line, data_integer);
14.     data_vec    <= conv_std_logic_vector(data_integer,8);
15.     clock20Mhz  <= data_vec(0);
16.     reset       <= Not data_vec(7);   -- inverted data bit 7 for reset signal
17.     input_pulse <= data_vec(6);
18.     Wait For 25 nsec;
19.  End Loop;
20. file_close(data_in);        -- closes the file once the loop has completed
21. End Process;
```

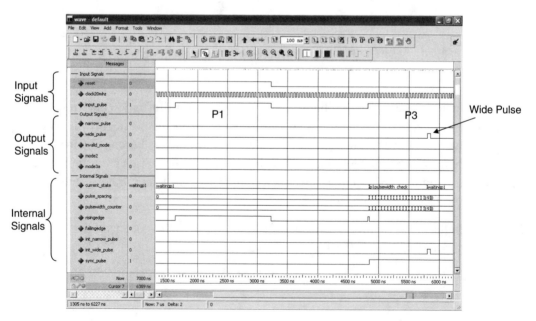

Figure 5-16: TLA Application Data Output Waveform (Material based on or adapted from figures and text owned by Xilinx, Inc., courtesy of Xilinx, Inc. Copyright © Xilinx 1995–2008 used in Xilinx ISE WebPack™ software version 10.1.)

Lines 3–21. Read External File Process

This process reads the data from the external data file and is inserted in the architecture following the initial Begin statement.

Lines 11–14. Assigning Data to Input Signals

First a line is read from TLA_Data.txt. Since each line has only one number, the readline contains one number. This number is converted to an 8-bit standard logic vector and assigned to data_vec. Only bits 0, 7, and 6 of data_vec are used.

Line 15. Assigning Clock's Input

Data_vec bit 0 is connected to clock20Mhz.

Line 16

Data_vec bit 7 is inverted and connected to reset.

Line 17

Data_vec bit 6 is connected to input_pulse.

5.7. Simulation Tutorial

This tutorial demonstrates how to set up and run a simulation using ModelSim XE III 6.4b, included in Xilinx ISE WebPack™. After completing this tutorial, it will be easy to apply the same process to other VHDL designs and testbenches. The same basic concepts apply when performing simulation on other simulators.

Simulation Assumptions

- Preinstalled ModelSim XE III 6.4b.

- Design code filename Mode2n3.vhd is located at C:\Chapter 5 Simulation \Design Code.

- Testbench filename tb_mode2n3.vhd is located at C:\Chapter 5 Simulation \Testbenches.

Invoke ModelSim

Select Start → All Programs → ModelSim XE III 6.4b → ModelSim or the icon on your desktop. Note: The path may be different depending on the operating system.

Create a New Project

Select File → New → Project.

Name the project setup, see Figure 5–17. Type Modes for the project name.

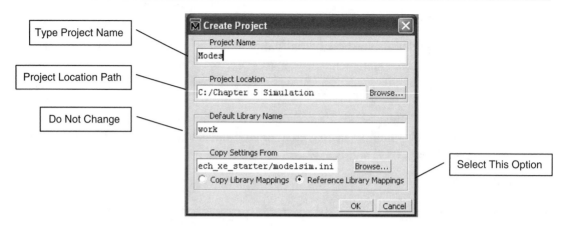

Figure 5–17: Create a Project (Material based on or adapted from figures and text owned by Xilinx, Inc., courtesy of Xilinx, Inc. Copyright © Xilinx 1995–2008 used in Xilinx ISE WebPack™ software version 10.1.)

Browse to `C:/Chapter 5 Simulation`.

Keep default library name `Work`. The complier automatically creates the default work library for the design code. This is where the compiled code is placed.

Select the `Reference Library Mappings` option under `Copy Settings From`. Note: Either option will work, I prefer just to map to the original. I learned the hard way that making copies can cause problems.

Select `OK`.

Add Files to the Project

The next pop-up window `Add items to the Project` (Figure 5–18) allows you to create a new file, create a simulation, create a new folder, or add an existing file. The design code and testbench already are written, so we are going to add those files to the project.

Double click on `Add Existing File`.

Add Design Code to the Project

Select `Browse` and navigate to design code located at `C:\Chapter 5 Simulation \Design Code` (Figure 5–19). Note: Make sure `Reference from current location` is selected.

Select `Mode2n3.vhd`.

Select `Open`.

Select `OK`.

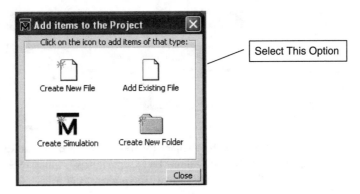

Figure 5-18: Add Item to the Project (Material based on or adapted from figures and text owned by Xilinx, Inc., courtesy of Xilinx, Inc. Copyright © Xilinx 1995–2008 used in Xilinx ISE WebPack™ software version 10.1.)

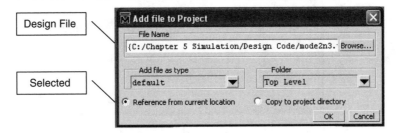

Figure 5-19: Adding Design Code to the Project (Material based on or adapted from figures and text owned by Xilinx, Inc., courtesy of Xilinx, Inc. Copyright Xilinx © 1995–2008 used in Xilinx ISE WebPack™ software version 10.1.)

Add Testbench Code to the Project

Double click on `Add Existing File` in the `Add items to project` window when it reappears.

Select `Browse` and navigate to `testbench code` located at `C:\Chapter 5 Simulation\Testbenches` (Figure 5–20). Note: Make sure `Reference from current location` is selected.

Select `tb_mode2n3.vhd`.

Select `Open`.

Select `OK`.

Note: Multiple files can be selected and added at the same time if they are located in same directory. Just hold down the control (`ctrl`) key before you select `Open`.

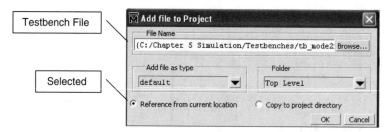

Figure 5–20: Adding Testbench Code to the Project (Material based on or adapted from figures and text owned by Xilinx, Inc., courtesy of Xilinx, Inc. Copyright © Xilinx 1995–2008 used in Xilinx ISE WebPack™ software version 10.1.)

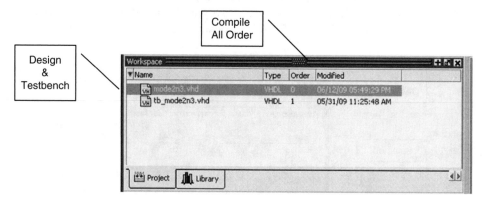

Figure 5–21: Project Files (Material based on or adapted from figures and text owned by Xilinx, Inc., courtesy of Xilinx, Inc. Copyright © Xilinx 1995–2008 used in Xilinx ISE WebPack™ software version 10.1.)

Select `Close` when the `Add Existing File` pop-up window reappears, since all files have been added.

The `Workspace` window shows the filenames of the code in the project, type, the order in which they will be compiled, and the last time the file was modified (Figure 5–21).

Compile Files

Files in the project can be compiled all at once or one at a time.

Option 1: Compile Selected File

This option compiles the selected file(s) highlighted in the `Workspace` window.

Select `mode2n3.vhd`.

Select `Compile` → `Compile Selected`.

Option 2: Compile All Files

This option compiles all the files in the `Workspace` window at once according to the order number.

Select `Compile → Compile All`.

First, the design code `mode2n3.vhd` is complied, since it has order #0, then the testbench `tb_mode2n3.vhd`. With this option it is not necessary to highlight files.

Feel free to try both ways.

Compiler Error

I created an error by removing the semicolon (`;`) from line 27 in the testbench code, shown in Figure 5–22. The transcript window shows total number files compiled and total number of successful and failed files.

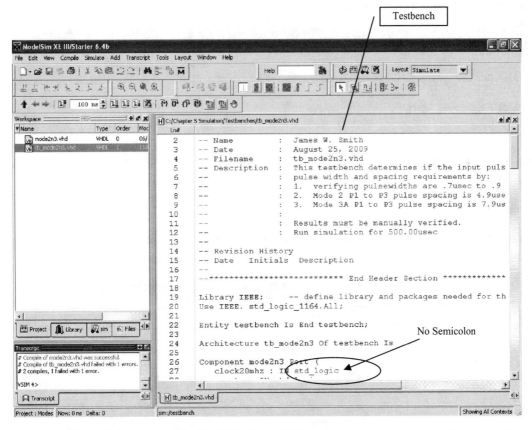

Figure 5–22: Failed Simulator File Compile (Material based on or adapted from figures and text owned by Xilinx, Inc., courtesy of Xilinx, Inc. Copyright © Xilinx 1995–2008 used in Xilinx ISE WebPack™ software version 10.1.)

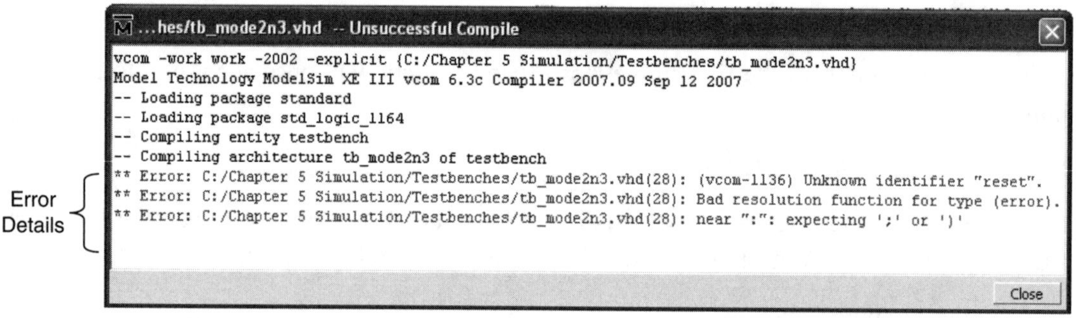

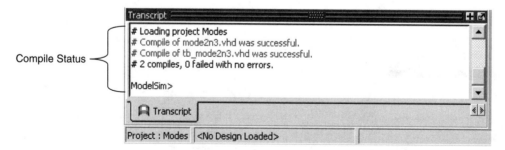

Figure 5-23: Simulator Compile Error Details (Material based on or adapted from figures and text owned by Xilinx, Inc., courtesy of Xilinx, Inc. Copyright © Xilinx 1995–2008 used in Xilinx ISE WebPack™ software version 10.1.)

Select `Compile → Compile Report` to view details about error, see Figure 5–23. After reviewing the report, go to the file to make the correction. Since this is minor, I make the correction using the ModelSim editor.

Double click on `tb_mode2n3.vhd` to open the viewing area window to the right. Add a semicolon (;) to line 27.

Save File Corrections

Select `File → Save`.

Recompile using either option 1 or option 2.

Now the file contains no error, and the transcript window shows the files were successfully complied, see Figure 5–24.

Simulate

Now that the design and testbench have been successfully compiled, it is time to start the simulation. Select `Simulate → Start Simulation` (Figure 5–25).

Figure 5-24: Successful Compile Status in Transcript Window (Material based on or adapted from figures and text owned by Xilinx, Inc., courtesy of Xilinx, Inc. Copyright © Xilinx 1995–2008 used in Xilinx ISE WebPack™ software version 10.1.)

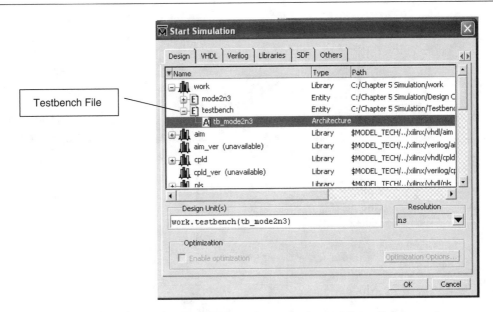

Testbench File

Figure 5–25: Select Testbench (Material based on or adapted from figures and text owned by Xilinx, Inc., courtesy of Xilinx, Inc. Copyright © Xilinx 1995–2008 used in Xilinx ISE WebPack™ software version 10.1.)

In the pop-up window, click on the + next to `work` then `testbench` to select `tb_mode2n3`. Note: If other files had `testbench` as their entity's name, their architecture's name would have also appeared. All my testbenches have the entity name of `testbench`. The reason I do this is because I will have several testbenches for one design. By having all their entities named testbench, they appear under the one entity, making it is easier for me to find the specific testbench. My architecture's name is descriptive to what the code is verifying.

Select `Resolution` to `nsec`.

Select `OK`.

The transcript window will let you know when the design has been loaded. Now you are ready for the waveform and signals.

Select `View → wave`.

Select all the signals shown in the `Objects` window. Make sure `testbench` is highlighted in the `Workspace` window (Figure 5–26), then right mouse click on the `Objects` window.

Select `Add to Wave → Selected Signals`; this adds the signals you selected to the `wave` window.

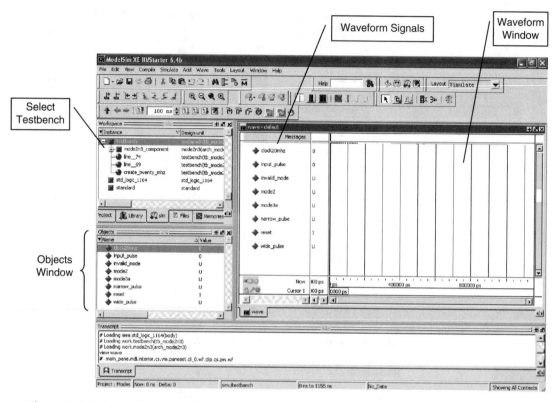

Figure 5–26: Load Simulator (Material based on or adapted from figures and text owned by Xilinx, Inc., courtesy of Xilinx, Inc. Copyright © Xilinx 1995–2008 used in Xilinx ISE WebPack™ software version 10.1.)

Click in the transcript window to move the cursor.

Type in `run 10µsec`.

Select `Enter`. The simulation output will be displayed in the waveform window, see Figure 5–27.

View Output Listing

To see the output as a listing, do the following.

Select `View → List` to open the `List` window.

Select all the signals shown in the `Objects` window.

Make sure `testbench` is highlighted in the `Workspace` window, then right mouse click in the `Objects` window.

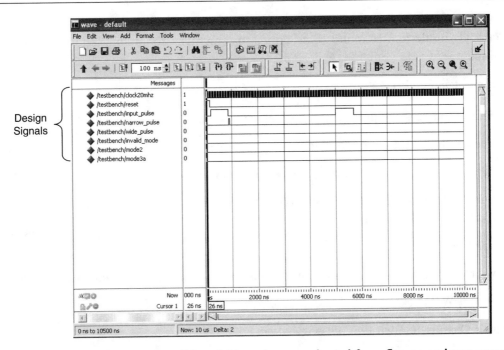

Figure 5–27: Waveform Output (Material based on or adapted from figures and text owned by Xilinx, Inc., courtesy of Xilinx, Inc. Copyright © Xilinx 1995–2008 used in Xilinx ISE WebPack™ software version 10.1.)

Select Add to List → Selected Signals. This will add the signals you selected to the wave window.

This tutorial has shown you how to perform a simulation using ModelSim III XE 6.4b. Remember, other simulators work differently, but the output for the same design should be the same. Now that you have some of the basics of how to perform a simulation, take some time to explore the many other features available using ModelSim III XE 6.4b or your simulator.

5.8. Chapter Overview

I find simulation to be the most enjoyable and exciting of the FPGA development phases—so many options are available for simulating an FPGA design. Depending on the situation, one option may be more beneficial than another. The next chapter covers synthesis. However, if you performed synthesis prior to simulation (and I hope you plan on simulating), then your simulation phase may include both RTL and functional simulations. If this is the case, then remember, for each design modification, a new postsynthesis netlist for simulation must be created and used for the most accurate functional simulation. Here are some things to remember about simulation as you continue to develop your FPGA design.

Key Simulation Phase Tips

- Simulation is not required but should always be performed, especially on new designs.

- Testbenches are a great reusable way to apply design stimulus.

- Write test cases to help develop testbench stimulus.

- Simulation allows a design to be verified under various test conditions and limits without damaging the hardware.

Chapter Links

For your convenience here are some links to a couple of complete development tools.

Xilinx ISE WebPack: www.xilinx.com/tools/designtools.htm.

Altera's Quartus II, Web Edition: www.altera.com.

More information on the offline logic analyzer or pattern generator, Tektronix: www.tek .com.

The Moving Pixel Company: www.movingpixel.com/main.pl?home.html.

CHAPTER 6

Synthesis

6.1. Introduction

Synthesis is the point in FGPA development where a high-level design is broken down into a mid-level netlist that is now associated with logic and internal FPGA resources. The design can be the one presented in this book or one you or someone else created or modified. It can be in several different formats—HDL, schematic capture, or a mixture—and may have been verified through simulation. In spite of whoever created or modified the design and the format, simulated or not, the design must be synthesized before it can be programmed into an FPGA.

Although, in this book, synthesis is performed following simulation, it could be performed immediately following the design phase. Once the design is complete it must somehow get broken down to a format that describes and connects the same functions in terms of FPGA resources. How do we make this happen? The answer is that the design must go through a two-step process: first synthesis and then implementation. These steps take the high-level design and break it down to a format that eventually gets programmed into an FPGA.

This chapter discusses the design synthesis phase or process. Synthesis is the first place in which the HDL design is associated with the internal logic. The input to the synthesis phase is the design, and the output consists of a design netlist that feeds into the implementation tool and an option for a functional simulation netlist, see Figure 6–1. Additional outputs include a report file and schematic views, which provide pertinent information about the synthesized design. These files are discussed later in this chapter. Unlike the simulation phase, which is optional but highly suggested, synthesis is mandatory; and synthesis must be performed before implementation.

For some designs, the synthesis process can be performed with much ease; however, for other designs, the process can be complex and long. Our ultimate goal is to create a netlist that connects the FPGA's resources to perform the same functions as defined by the high- level design. The first step to accomplish this is the synthesis phase. As we continue down the FPGA development path, we get closer to having a design that can be programmed into a device. While synthesis may not be as exciting as simulation (at least in my opinion), it is required and can be time consuming.

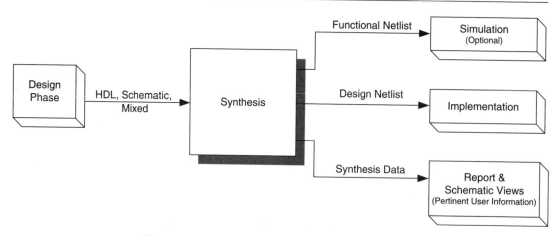

Figure 6–1: Synthesis Phase Inputs and Outputs

In this chapter, you will learn

- The design synthesis process.

- Synthesis tools and manufacturers.

- Synthesized output files.

- How to perform synthesis, through a tutorial.

6.2. What Is Design Synthesis?

The FPGA device consists of logic blocks or cells that are configured to perform the functions defined by the high-level design. So far, all we have is a high-level design but nothing that associates it with the internal FPGA resources. Design synthesis or synthesis is the process that takes the high-level design associates it with FPGA resource and reduces logic to make the design more efficient. It can best be described as a three-step process that converts a high-level design to a mid-level design netlist, see Figure 6–2. The reason I say *mid-level design netlist* is because it cannot be used to program an FPGA, but it is just one development stage from being ready to burn into a chip. Synthesis is the first step in the development process in which the design is associated with the FPGA's internal logic technology. In other words, the output netlist is a little more realistic because the device's part number is defined and available resources are known and used to create the netlist that has some timing information.

The three basic synthesis operations (Figure 6–3) are syntax check and element association, optimization, and technology mapping. Generic synthesis operation terms are used to distinguish one step from another. Each synthesis tool may call the steps something different,

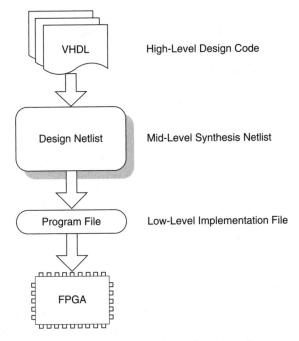

Figure 6–2: Mid-Level Synthesis Netlist

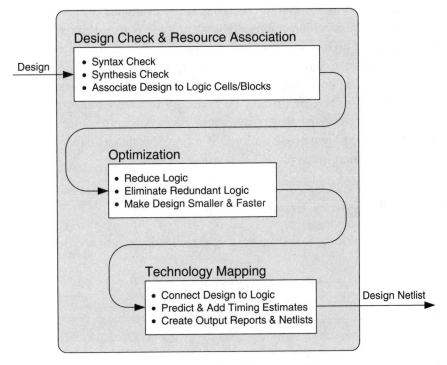

Figure 6–3: Synthesis Process Flow

but they perform the same basic functions. Third party tools generate an output netlist that can be imported into an implementation tool and simulator.

6.2.1. Design Check and Resource Association

First, the design is checked for syntax and synthesis errors. Nonsynthesizable command errors do not cause synthesis errors. For example, the `after` command, which was used in the testbench, creates a delay and is used by the simulator. It is not synthesizable and is ignored by the synthesis tool. Missing or misplaced semicolons or misspelled keywords will cause the synthesis tool to generate an error, see Example 6–1.

Once the design is error free, it is converted into structural elements. This means that logic elements are inserted as replacements for things like an addition sign (+), subtraction sign (–), or for inferred flip-flops, gates, registers, and the like.

6.2.2. Optimization

At this point, the design is represented by interconnecting the internal FPGA resources to mimic the functions defined by the high-level design. In this state, the design is just put together without concern for redundant logic, timing constraints (if provided), clock speed, and other design considerations. Now that the design is put together, algorithms are used to optimize the design. This means that the design is really examined for things like redundant logic, clock speed, and timing constraints. Redundant logic is removed to make the design smaller. Algorithms are used to evaluate multiple paths to ensure the fastest timing is achieved. The shortest routing distance does not necessarily mean the fastest time. Because of the resource layout and how those resources are used, the shortest distance may not produce the fastest time. Therefore, it may be necessary to have a longer route to meet timing requirements, because the shorter route may require more resources, resulting in longer time. As shown in Figure 6–4, option 2 is a longer distance, however, option 1 has more resource delays; therefore, option 2 is the faster route.

■ Example 6–1. Syntax Error

The first signal definition causes the synthesis tool to generate a syntax error message:

```
Signal pulse : std_logic    -- missing semicolon (;) following
std_logic, error message generated
Signal pulse : std_logic;    -- no error message generated
```

■

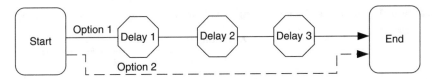

Figure 6–4: Faster Routing Path

6.2.3. Technology Mapping

Now that the design has been optimized, it is mapped to the technology associated with the targeted FPGA. Information such as the FPGA part number, speed and manufacturer is provided when setting up the synthesis tool. Examples of some technology view symbols are shown in Figure 6–5. Synthesis tools use advanced techniques to make predictions about how the design will be place and routed in the target device. These advanced techniques produce synthesis timing estimates that are near the actual postimplementation timing. However, the real timing is unknown until after the design has been placed and routed.

6.3. Synthesis Phase Tools

The tools needed for the synthesis phase are a synthesis tool, or synthesizer, and an editor to modify the original design if necessary. Because an HDL text editor is included with many synthesis tools, you could use this for HDL editing. My personal preference is my original text editor, but for small changes, such as correcting synthesis or syntax errors, I generally use the synthesis editor. If you decided to switch between the two editors, make sure that the changes have been applied to both copies of the design. Synthesis tools are available as standalone or part of a complete package. Some of the advantages and disadvantages to standalone tools versus complete package tools are listed in Tables 6–1 and 6–2.

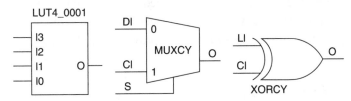

Figure 6–5: Technology View Symbols (Material based on or adapted from figures and text owned by Xilinx, Inc., courtesy of Xilinx, Inc. Copyright © Xilinx 1995–2008 used in Xilinx ISE WebPack™ software version 10.1.)

Table 6–1: Complete Package Synthesis Tool Advantages and Disadvantages

Advantages	
Single tool	Need to know only one tool
Faster process	Eliminates time to switch between third party tool(s)
Cost	Single tool may be cheaper than multiple tools
Expert on device	Manufacturer understands device better than a third party Device data are more accurate
Disadvantages	
Manufacturer dependent	Can't use synthesize netlist with other manufacturers
Synthesis netlist	Synthesis netlist may not be as good as a third party's May not utilize internal resources either
Supports only one manufacturer	Must obtain another tool for other manufacturers
Not area of expertise	Expert on device not necessarily on synthesis development

Table 6–2: Standalone Synthesis Tool Advantages and Disadvantages

Advantages	
Manufacturer independent	Supports multiple manufacturers Easy to switch vendors Output netlist available in different manufacturer formats
Area of expertise	Synthesize netlist generally better than manufacturer's May provide better synthesized netlist
Disadvantages	
Multiple tools	Separate tools for synthesis and implementation
Cost	May be more expensive than complete tool
Not expert on device	Manufacturer understands more about device than a third party Estimated device timing data may not be as good as manufacturer's

As a result of the continuous evolution of FPGA gate count from hundreds, to thousands, to millions of gates and increasing functionality, the synthesis tools have evolved as well. New, more advanced FPGA features led to newer, more advanced tools. In the past, there were few choices for synthesis tools, and many companies offered only one synthesis tool. Now, more companies offer a selection of synthesis tools, each providing slightly more or different features.

6.3.1. Vendors and Features

Today, many more options for synthesis tools are available than years ago. Not only do many manufacturers make the tools, it is becoming common for standalone manufacturers to offer different levels or features in their synthesis tools.

This section provides information on complete package and standalone synthesis tools and some of their features. Synthesis is performed as a part of Altera's Quartus II® and Xilinx ISE® complete development packages. Many manufacturers make claims to have the best, world's first, or some other claim about their synthesis tools. My opinion is that, depending on your design, some tools perform better than others, but you have to decide for yourself. Some commonality among most synthesis tools includes

- Allow user to perform syntax check only.

- Create RTL view.

- Create technology view.

- Generate synthesized netlist.

- Generate functional simulation netlist.

Quartus II® offers users two synthesis options "Analysis and Elaboration" and "Analysis and Synthesis." Where analysis and elaboration is just a presynthesis step that

- Performs syntax and semantic error checks.

- Does not perform logic synthesis or technology mapping.

The complete synthesis process is performed by the analysis and synthesis option that

- Checks for syntax and semantic errors.

- Minimizes design logic.

- Performs technology mapping.

ISE Design Suite by

- Xilinx Synthesis Technology (XST)

  - Incorporates next-generation physical synthesis optimizations by using techniques such as register balancing, global optimization, timing-driven synthesis, and logic optimization.

  - Provides reduced runtime and design preservation.

  - Reduces power use by using power-aware optimization.

  - Provides integrated RTL and Technology. Viewers to view the RTL netlist.

Keep in mind that Quartus II supports only Altera's FPGA devices and ISE supports only Xilinx. On the other hand, standalone packages, such as the ones offered by Mentor Graphics and Synopsys, are vendor independent.

Mentor Graphics offers LeonardoSpectrum®, Precision RTL®, Precision Physical®, and Precision RTL Plus® synthesis tools. The Precision tool sets offer a progressive line of features.

LeonardoSpectrum offers

- F.A.S.T. optimization. Proprietary algorithm with high quality of results (QoR).

- Incremental synthesis. Reduces compile time for multiple or large designs.

- Partitioning. Makes it easy to divide or partition designs.

Precision RTL offers basic features such as

- Advanced Optimization Algorithms.

  - Maximum use of FPGA resources.

- RTL and Technology Viewers.

- Interactive Static Timing Analysis.

- DSP and RAM Inference Optimization.

- Gated Clock Conversions.

- Register Retiming.

Precision RTL Plus offers the same features as Precision RTL and

- Physically Aware Synthesis.

  - Optimizes based on preimplementation estimates and considerations such as delays, potential placement, routing, and other device-related design rules.

- Incremental Design Flows.

  - Can recompile and synthesize portions of the design.

- Resource Manager.

  - Interface that allows the designer to analyze and manipulate the mapping of the FPGA's resources to optimize performance or area.

Synopsys offers Synplify Pro and Premier.

Synplify Pro uses

- Behavior extracting synthesis technology (B.E.S.T.) optimizer for

  - Proprietary algorithms.

  - HDL Analyst.

- Creates an RTL block diagram.

- Graphical state machine viewer.

- Automatic RAM and DSP inference.

- Incremental Design.

- Automatic Retiming.

  - Improves timing performance and balances delays by moving registers within combinatorial logic.
- FSM Compiler and Explorer.

  - Optimizes finite state machines based on constraints

Synplify Premier provides some of the same features as Synplify Pro but also some extras

- A simulator–like debug environment.

- DesignWare compatible library.

- Easy ASIC code migration.

- SynCore IP wizard.

- Automatically generates technology-independent RTL for memories and first-in/first-outs.

6.3.2. Synthesis Tool Setup

Before performing synthesis, there is a little tool setup. Until this point in the development phase, the FPGA's part number was unnecessary; however, the synthesis process needs information about the part, such as speed and available internal resources, to synthesize the design. The FPGA is identified by selecting the family, device number, package, and speed. Any of those selections is easy to change and resynthesize the design using the new information. Using a third party tool also makes it easy to switch between different manufacturers. Basic tool setup consists of creating a project that contains all the information about the design, see Figure 6–6. Some general information provided during the project setup includes

- Device information (i.e., family, device number, package, and speed), which may be found on the device package.

- Input design.

- User-defined constraint file(s).

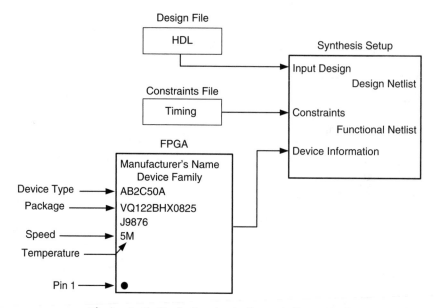

Figure 6–6: Basic Synthesis Setup Information

6.4. Synthesis Input

The input to the synthesis development phase is the design, VHDL code in our case. In addition to the design file are the user-defined constraints or limitations. The constraints may contain such things as timing or Vendor attributes.

Acceptable design formats vary between synthesis tools, so make sure prior to creating your design or selecting your tool that the formats are compatible.

Altera's Quartus II accepts

- AHDL (Altera Hardware Description Language).

- VHDL.

- Verilog.

- System Verilog.

- Schematic capture.

- EDIF input files; Quartus II supports both .edif and .edn file extensions.

- Verilog Quartus mapping files; this is a node-level netlist in ASCII text format, generally created by an EDA synthesis tool, like Synopsys Synplify.

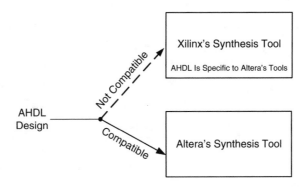

Figure 6–7: Design Format and Synthesis Tool Compatibility

Xilinx's Synthesis Technology accepts VHDL and Verilog.

Mentor Graphics's LeonardoSpectrum® accepts a mixture of VHDL, Verilog. and EDIF.

Precision RTL, Precision Physical, and Precision RTL Plus Synthesis accepts Supports System Verilog, Verilog, VHDL, EDIF, or a combination of these.

Synopsys's Synplify Pro accepts VHDL, Verilog, and a mixture of VHDL and Verilog. So if your synthesis tool is Synplify, your design format cannot be schematic capture or AHDL, which is compatible with Altera's development tools, see Figure 6–7. For most cases, especially if you use mainstream tools and HDL, there is no problem with compatibility. However, just as a sanity check, it is a good idea, when you first start, to make sure that no compatibility issues exist.

6.5. Synthesis Output Files

The synthesis process generates many different types of files. Some will be used during the FPGA development process and others will have no meaning to you, see Figure 6–8.
The synthesis process has done a lot of work to break down the high-level design to a lower level. At the completion of the synthesis phase the original design is closer to a format that will be used to program an FPGA.

Netlists, status reports and schematic views are some of the outputs generated by the synthesis tool that will help you during the development process.

* Netlists: a design netlist, which is the synthesized design, and an optional functional netlist, used to perform functional simulation.

* Status reports that states internal resources utilization, critical timing path(s), and other pertinent information.

* Schematic views: RTL and technology.

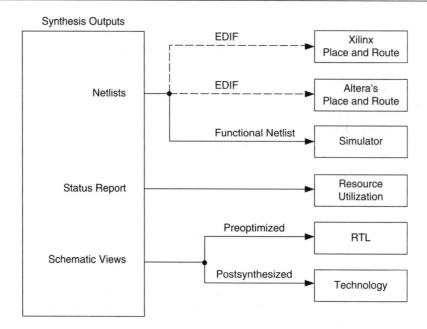

Figure 6–8: Synthesis Outputs

6.5.1. Netlists

The design netlist is what the original design looks like after it has been synthesized (i.e., optimized, connected using internal FPGA logic). It is not necessary for the complete development tools to output the design netlist, since synthesis and implementation are combined in one tool. Third party tools produce this netlist, so it can be used as the input to the manufacturers' implementation tool. Consult the implementation tool to determine which file extension(s) are acceptable. Most implementation tools accept the generic EDIF format and their manufacturer-specific format. Even though the design is represented by this netlist, it cannot be used for simulation. Most synthesis tools provide an option to have a functional simulation netlist or file generated. Generally, the functional simulation file is not generated automatically. I suggest generating this file and performing the functional simulation, especially if time permits. The functional simulation allows you to verify that the synthesis process did not change the design. You should be able to use the RTL testbench to verify the netlist; and since this netlist represents the original design, you should expect the same results. In the RTL simulation, results were instant because no timing delays were used, but the synthesis tool introduces timing delays, which may be viewable during your simulation.

The functional netlist is complied in the simulator instead of the original design and verified using a testbench. If the netlist requires additional libraries like the unisim library, it must be added to the simulator's library. The design should perform the same way, proving the synthesis process has not changed the design. If this isn't the case, then it will be necessary to utilize different synthesis tool features and files to determine the problem(s).

6.5.2. Status Reports

Now that the design is interconnected and utilizes internal resources, we can now know how much of the internal resources are used, clock and other timing information, critical paths, warnings, errors, and we can even see the design represented as a schematic. These status report files are output from the synthesis tool and are not used as input to other development phases. Their main goal is to provide the user with helpful information about the design and allow him or her to identify real and potential problems, such as the design not meeting timing and other constraints. Depending on the design complexity, you may review one or all of the output files. Each synthesis tool provides information in different ways, so consult the user's manual to determine what files contain this information.

6.5.3. Schematic Views

The synthesis tool generates two schematic views: RTL and technology. The RTL schematic view shows the preoptimized design in terms of generic symbols, such as adders, multipliers, counters, AND gates, and OR gates. This view is manufacturer independent. One of the main benefits of this view is that some design issues may be detected by viewing the RTL schematic and corrected early in the development process. You can think of this view as being raw, in that it is not associated with a manufacture, because nothing has been done to reduce the logic. It has just been translated from the high-level design.

6.5.4. Technology Schematic View

After the design has been synthesized, it can be viewed as a schematic, which is represented by the technology schematic view. This view shows gates and elements as they will look in the selected manufacturers' device. Now, the design looks more like it will when it is put into the FPGA. You should review this schematic to make sure that the synthesis process has not removed logic you wanted in the design. If this happens, it may be necessary to rewrite some code or use constraints to keep the logic.

The output files provide a good first look at resource utilization and timing, so design modifications can be made prior to implementation. Because of the many aspects of performing synthesis, I think it would really help to show an example of how synthesis is performed. In the next section, a synthesis tutorial is provided using Xilinx's XST synthesis tools. Xilinx's Webpack ISE 10.1®, which provides synthesis, implementation, and a simulator, has been downloaded for the tutorial. The best part is it is *free*. I suggest that, if you have no development tools, this would be a good one to download and try. The Webpack ISE is available for both Windows and Linux and can be found at www.xilinx.com/tools/webpack .htm. A free user account is required, so you have to create a user name and password before getting access.

6.6. Synthesis Tutorial

Xilinx's XST synthesis tool is a part of its ISE and Webpack ISE complete development packages. Whereas ISE is a fee-based tool that supports more devices and offers more features, the Webpack is free, with limited features but sufficient for this tutorial. Keep in mind that your synthesis tool may require a different setup and the terminology may be slightly different, but in the end, a third party tool provides a synthesized netlist that will be used in the implementation development phase and an option to output a functional simulation netlist. It is unnecessary for the manufacturers to provide the synthesized netlist, and this may not be provided.

Launch Synthesis Tool

Select `Start → All Programs → Xilinx ISE Design Suite 10.1 → ISE → Project Navigator` or click on the desktop icon. Note: Depending on your operating system, your `Start` path may be slightly different.

Create New Project

Select `File → New Project`. A "New Project" wizard is provided to step you through creating a project. The project name, location, and top-level source type are defined as

> *Project name:* `Mode2n3_project`
>
> *Project location:* `C:\Chapter6_Synthesis\Mode2n3_project`
>
> *Top-level source type:* `HDL`

Note: See Figure 6–9 for selections in the project wizard.

Select `Next`.

Now it is time to tell the synthesis tool which FPGA device will be used, the format of your design format, and your selected simulation tools.

Device Properties

Select `General Purpose`.

Not all families are available for each product category, each family offers different devices, and not all devices have the same packages or speed, so depending on your selection, the pull-down menu options will vary.

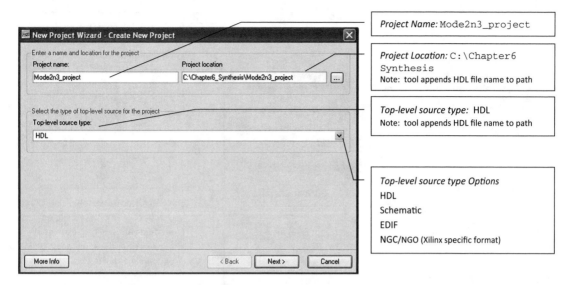

Figure 6–9: Create New Project (Material based on or adapted from figures and text owned by Xilinx, Inc., courtesy of Xilinx, Inc. Copyright Xilinx © 1995–2008 used in Xilinx ISE WebPack™ software version 10.1.)

Select `Family` → `Spartan3`.

Select `Device` → `XC3S50`.

Select `Package` → `VQ100`.

Select `Speed` → `–5`.

Select `Top-Level Source Type` → `HDL`.

Select `Synthesis Tool` → `XST` (VHDL/Verilog).

Note: At one time, some manufacturers offered third party synthesis tools to allow you to compare synthesized netlists. If this were the case, then the `Synthesis Tool` pulldown menu would provide options. No other synthesizers are provided as an option. See Figure 6–10 for selections in the project wizard.

Select `Simulator` → `Modelsim XE VHDL`.

Select `Preferred Language` → `VHDL`.

Keep default for the rest of the selections.

Select `Next`.

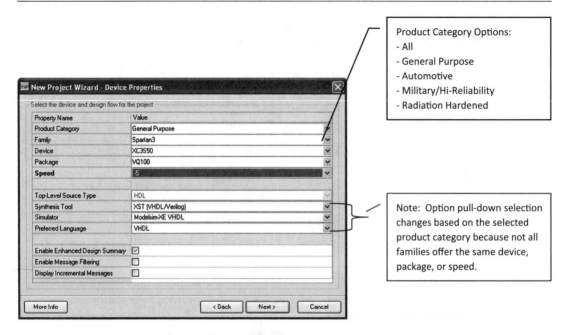

Figure 6–10: Device Properties (Material based on or adapted from figures and text owned by Xilinx, Inc., courtesy of Xilinx, Inc. Copyright Xilinx © 1995–2008 used in Xilinx ISE WebPack™ software version 10.1.)

Create New Source

The next screen, Create New Source, shown in Figure 6–11, allows for a new source file to be created; but since a design file is already created, no new file will be created.

Select Next.

Add Existing Source File

Now the design file will be added to the project, see Figure 6–12.

Select Add Source and navigate to the location of the source file.

Deselect Copy to Project. I found out the hard way that it is best to keep only one design code copy.

Select Next.

Project Summary

The next screen is a summary of the project information entered in the previous screens, see Figure 6–13. Look over the information for completeness and correctness. Select Back if changes are needed. Otherwise select Finish.

Select Finish.

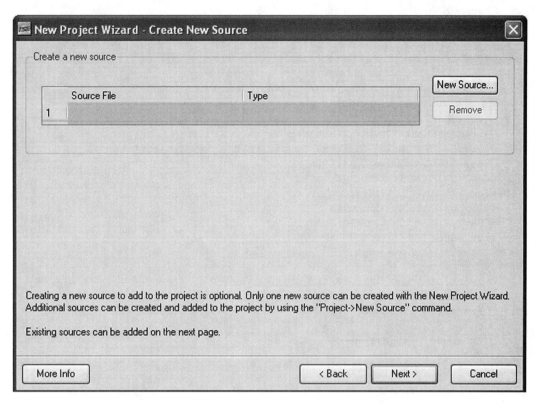

Figure 6–11: Create New Source File (Material based on or adapted from figures and text owned by Xilinx, Inc., courtesy of Xilinx, Inc. Copyright Xilinx © 1995–2008 used in Xilinx ISE WebPack™ software version 10.1.)

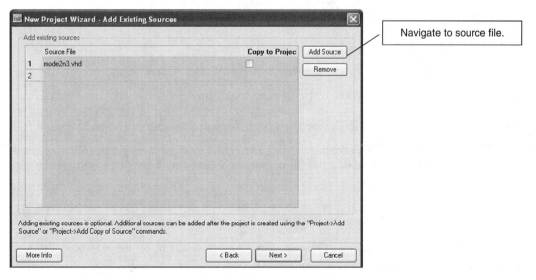

Figure 6–12: Add Existing Design File (Material based on or adapted from figures and text owned by Xilinx, Inc., courtesy of Xilinx, Inc. Copyright Xilinx © 1995–2008 used in Xilinx ISE WebPack™ software version 10.1.)

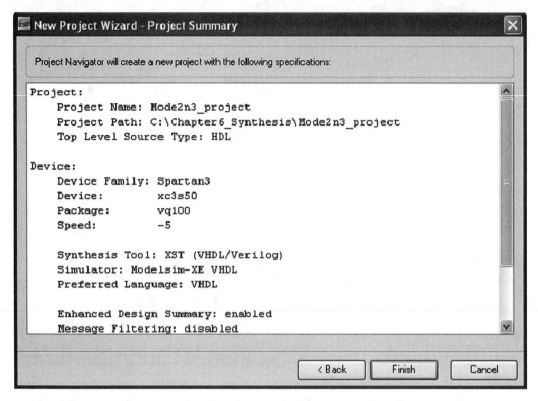

Figure 6–13: Project Summary (Material based on or adapted from figures and text owned by Xilinx, Inc., courtesy of Xilinx, Inc. Copyright © Xilinx 1995–2008 used in Xilinx ISE WebPack™ software version 10.1.)

Source File Status

This last screen allows you to view the status of the project source file(s).

Select `Association → All`. See Figure 6–14 for all the `Association` options.

Now that you have completed the wizard, you get the project window shown in Figure 6–15. This view has `Sources`, `Processes`, and a `transcript` section as well as a `viewing area`, where reports, source code, and other project information are displayed.

Sources Section

The sources section shows the project name, source file, and device part number.

Select `Sources for: → Implementation`. Double clicking on the source code opens the file in the viewing window, see Figure 6–16.

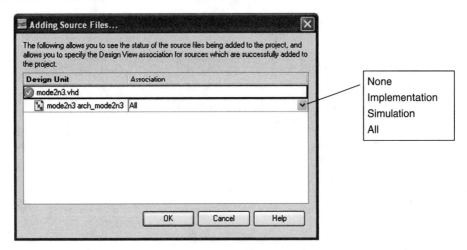

Figure 6-14: Source File Status (Material based on or adapted from figures and text owned by Xilinx, Inc., courtesy of Xilinx, Inc. Copyright © Xilinx 1995–2008 used in Xilinx ISE WebPack™ software version 10.1.)

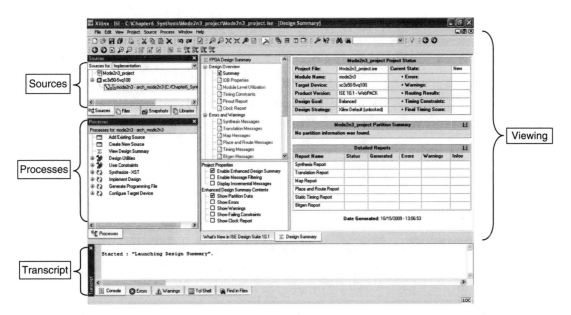

Figure 6-15: Project Main View (Material based on or adapted from figures and text owned by Xilinx, Inc., courtesy of Xilinx, Inc. Copyright © Xilinx 1995–2008 used in Xilinx ISE WebPack™ software version 10.1.)

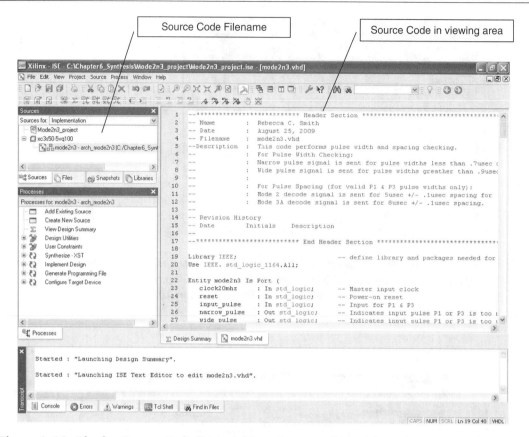

Figure 6–16: Viewing Source Code (Material based on or adapted from figures and text owned by Xilinx, Inc., courtesy of Xilinx, Inc. Copyright © Xilinx 1995–2008 used in Xilinx ISE WebPack™ software version 10.1.)

Transcript Window

The transcript window shows the status and provides other information about the processes. It contains several tabs that show information related to the named tab.

Processes Section

The processes section is where synthesis and implementation are performed; constraint, design summary, source and bit stream files are created; and you program the FPGA, but the one we are interested in for this chapter is the Synthesis -XST.

Expanding Synthesis XST in Process Window

Click on the + next to the Synthesis - XST. It shows the reports and schematic views generated by the synthesis tool, which are viewable by double clicking on the name. "Check syntax" (only checks the design's syntax) and "generate postsynthesis simulation modeling/netlist processing" are also performed here, see Figure 6–17.

Synthesize Design

Select `Synthesis - XST` and right click to see the synthesis options, see Figure 6–18.

Select Run.

Note: As the synthesis is running, the transcript provides status information; see Figure 6–19 for a sample. Errors detected during synthesis are displayed with error message(s) in

- The transcript window, see Figure 6–20.

- The process window with an X next to the `Synthesis - XST`, see Figure 6–21.

- The synthesis report file, see Figure 6–22.

- The design summary file, see Figure 6–23.

If the synthesis is successful (i.e., no errors are detected), then the transcript window will look similar to Figure 6–24 and the process and design summary like Figure 6–25.

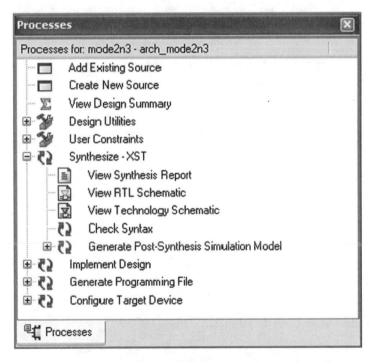

Figure 6–17: XST Expanded View (Material based on or adapted from figures and text owned by Xilinx, Inc., courtesy of Xilinx, Inc. Copyright © Xilinx 1995–2008 used in Xilinx ISE WebPack™ software version 10.1.)

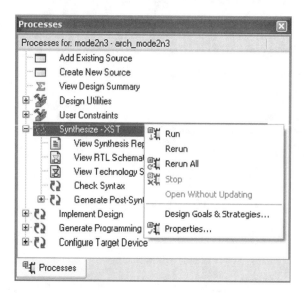

Figure 6-18: Start Synthesis (Material based on or adapted from figures and text owned by Xilinx, Inc., courtesy of Xilinx, Inc. Copyright © Xilinx 1995–2008 used in Xilinx ISE WebPack™ software version 10.1.)

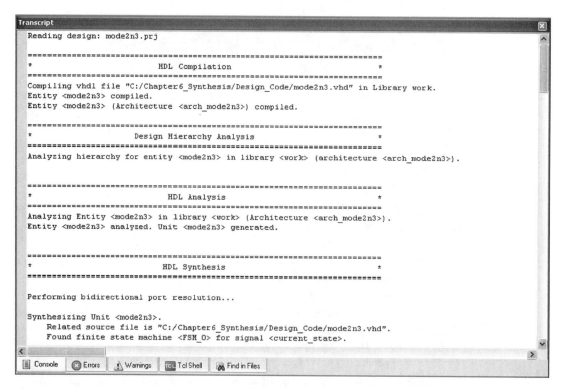

Figure 6-19: Synthesis Status in Transcript Window (Material based on or adapted from figures and text owned by Xilinx, Inc., courtesy of Xilinx, Inc. Copyright © Xilinx 1995–2008 used in Xilinx ISE WebPack™ software version 10.1.)

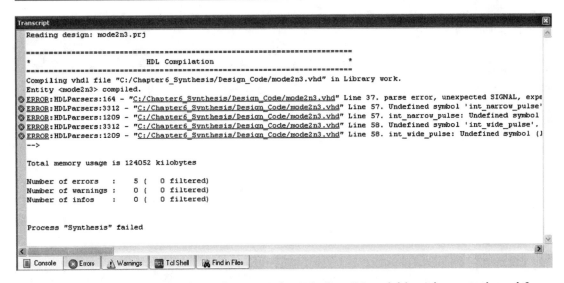

Figure 6–20: Synthesis Error Shown in Transcript Window (Material based on or adapted from figures and text owned by Xilinx, Inc., courtesy of Xilinx, Inc. Copyright © Xilinx 1995–2008 used in Xilinx ISE WebPack™ software version 10.1.)

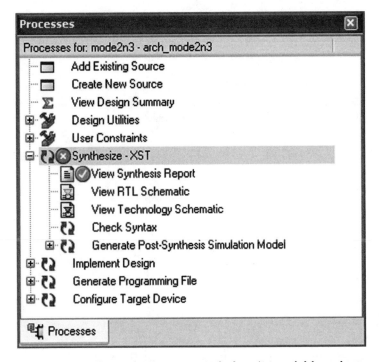

Figure 6–21: Synthesis Error Shown in Processes Window (Material based on or adapted from figures and text owned by Xilinx, Inc., courtesy of Xilinx, Inc. Copyright © Xilinx 1995–2008 used in Xilinx ISE WebPack™ software version 10.1.)

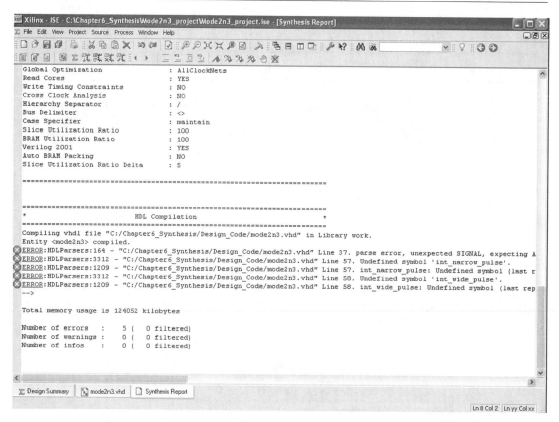

Figure 6–22: Synthesis Report File Shows Errors (Material based on or adapted from figures and text owned by Xilinx, Inc., courtesy of Xilinx, Inc. Copyright © Xilinx 1995–2008 used in Xilinx ISE WebPack™ software version 10.1.)

Now that the synthesis process is complete, the report and schematic views are available. A lot of information is available in the synthesis report, so I copied some information from the utilization summary.

Just for comparison the design was resynthesized using

Family: Spartan2

Device: XC2S15

Package: CS144

Speed: -5

Table 6–3 shows the synthesis comparison between the two devices.

The biggest differences between the two devices are the percent utilization of the internal resources and max frequency. Spartan 3 uses 20% of available slices while Spartan 2 uses

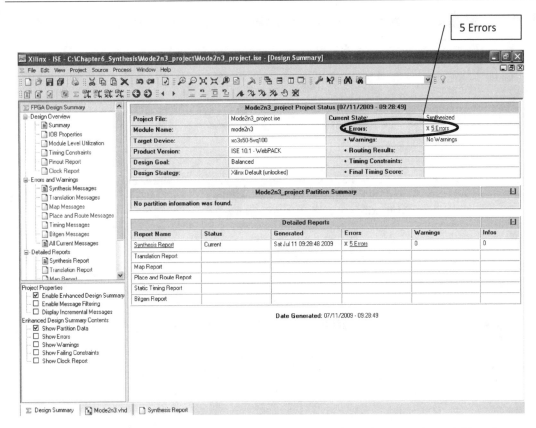

Figure 6–23: Design Summary File with Errors (Material based on or adapted from figures and text owned by Xilinx, Inc., courtesy of Xilinx, Inc. Copyright © Xilinx 1995–2008 used in Xilinx ISE WebPack™ software version 10.1.)

84%, and Spartan 3 has a maximum frequency of 112.3 MHz while Spartan 2 is 62.956 MHz. Because the Spartan 3 is a larger FPGA with more resources, its percentage of used slices is much less than Spartan 2. Both devices use about the same number of internal resources, which should be expected, since the design is the same.

It is important to derate how many internal resources are used. So I would think twice about using the Spartan 2 because 84% is a pretty high number for utilization. Some companies have a standard for derating the internal resources. Most of the time 60–70% is a good range. I generally like 50%, especially for new designs, where there is a good possibility that the design will grow. Utilizing a lot of your resources makes it difficult for the tool to synthesize and implement the design. Always leave yourself some growing room. Additionally, it is always a good idea to derate input/output pins to accommodate potential growth.

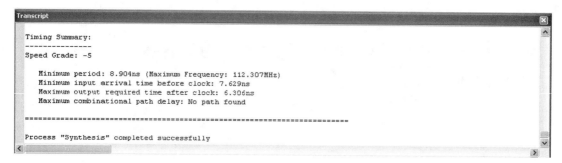

Figure 6–24: Successful Synthesis Shown in Transcript Window (Material based on or adapted from figures and text owned by Xilinx, Inc., courtesy of Xilinx, Inc. Copyright © Xilinx 1995–2008 used in Xilinx ISE WebPack™ software version 10.1.)

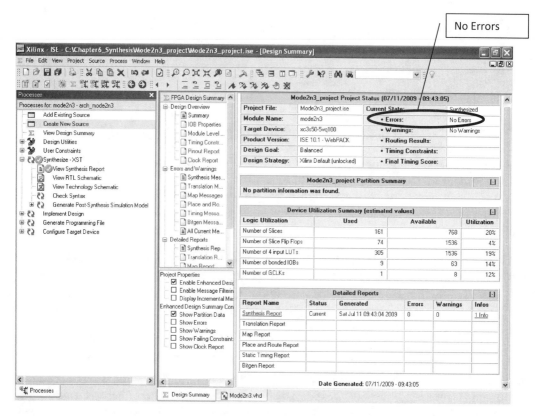

Figure 6–25: Successful Synthesis Process and Design Summary Windows (Material based on or adapted from figures and text owned by Xilinx, Inc., courtesy of Xilinx, Inc. Copyright © Xilinx 1995–2008 used in Xilinx ISE WebPack™ software version 10.1.)

Table 6–3: Spartan 3 versus Spartan 2 Utilization Comparison

Device Family	Spartan 3	Spartan 2
Part number	3s50vq100-5	2s15cs144-5
Number of slices	161 out of 768; 20%	162 out of 192; 84%
Number of slice flip-flops	74 out of 1536; 4%	74 out of 384; 19%
Number of 4 input LUTs	305 out of 1536; 19%	308 out of 384; 80%
Number of I/Os	9	9
Number of bonded IOBs	9 out of 63; 14%	9 out of 86; 10%
Number of GCLKs	1 out of 8; 12%	1 out of 4; 25%
Minimum period	8.904 nsec (Max Freq = 112.307 MHz)	15.884 nsec (Max Freq = 62.956 MHz)
Minimum input arrival time before clock	7.629 nsec	13.479 nsec
Maximum output required time after clock	6.306 nsec	8.329 nsec
Maximum combinational path delay	No path found	No path found

RTL View

The RTL schematic view shows how the design looks as it was converted to logic elements. Double click on `RTL Schematic` to see full view (Figure 6–26). The expanded view is shown in Figure 6–27.

Technology View

Double click on `View Technology Schematic`. This shows the internal technology, such as lookup tables connected to create the design. See a full view in Figure 6–28 and an expanded view in Figure 6–29.

Now that the design has been successfully synthesized, the optional functional simulation netlist can be created.

Create Optional Functional Simulation Netlist

This section creates the functional simulation netlist used to verify that the design was not changed as a result of the synthesis process. If the netlist is successfully created, checkmarks appear next to `Generate Post-Synthesis Simulation Model` and `Post-Synthesis Simulation Model Report`. The path and filename for the netlist is provided in the report file.

Select `Generate Post-Synthesis Simulation Model`. Then right mouse click to show the options, see Figure 6–30.

Selecting properties will show additional options, see Figure 6–31, but for this tutorial, the default values are good enough.
Select `Run`.

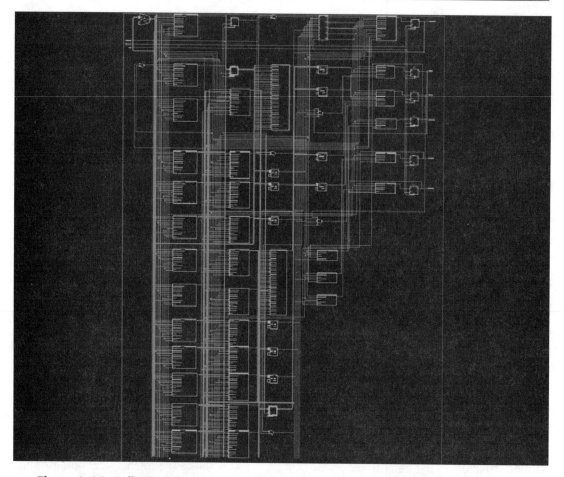

Figure 6–26: Full RTL Schematic View (Material based on or adapted from figures and text owned by Xilinx, Inc., courtesy of Xilinx, Inc. Copyright © Xilinx 1995–2008 used in Xilinx ISE WebPack™ software version 10.1.)

The netlist is in VHDL format but looks very different from the high-level design. The input and output ports are the same as in the design code, making it possible to use the same testbench. The netlist is a large file (about 5595 lines), so only samples of certain sections in the design are shown in Example 6–2. Note: format convention is different from the original design code.

Lines 1–5 is the library declaration.

Lines 7–19 is the entity section.

Line 21 starts the architecture section, only a small portion is shown.

Lines 22–25, shows some of the signal definitions.

Line 27 is the Begin statement for the architecture section.

Lines 28–41 shows some of the component instantiations.

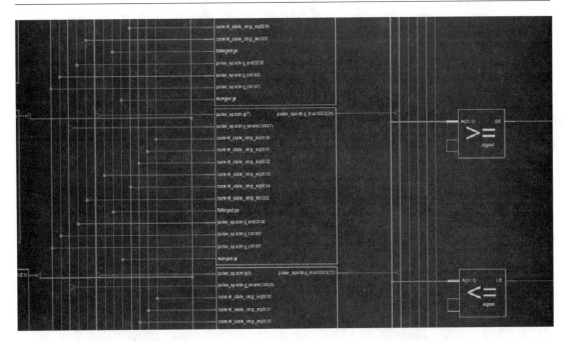

Figure 6–27: Zoomed in RTL Schematic View (Material based on or adapted from figures and text owned by Xilinx, Inc., courtesy of Xilinx, Inc. Copyright © Xilinx 1995–2008 used in Xilinx ISE WebPack™ software version 10.1.)

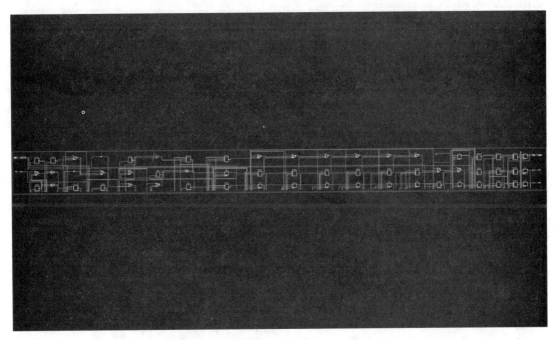

Figure 6–28: Full Technology Schematic View (Material based on or adapted from figures and text owned by Xilinx, Inc., courtesy of Xilinx, Inc. Copyright © Xilinx 1995–2008 used in Xilinx ISE WebPack™ software version 10.1.)

■ Example 6–2. Functional Netlist

```
1. library IEEE;
2. use IEEE.STD_LOGIC_1164.ALL;
3. library UNISIM;
4. use UNISIM.VCOMPONENTS.ALL;
5. use UNISIM.VPKG.ALL;
6.
7. entity mode2n3 is
8. port (
9.    reset : in STD_LOGIC := 'X';
10.       valid_pulse : out STD_LOGIC;
11.       mode3A : out STD_LOGIC;
12.       wide_pulse : out STD_LOGIC;
13.       input_pulse : in STD_LOGIC := 'X';
14.       mode2 : out STD_LOGIC;
15.       clock20Mhz : in STD_LOGIC := 'X';
16.       narrow_pulse : out STD_LOGIC;
17.       invalid_mode : out STD_LOGIC
18.    );
19.    end mode2n3;
20.
21.    architecture Structure of mode2n3 is
22.    signal Madd_pulse_spacing_share0000_cy_10_rt_2 : STD_LOGIC;
23.    signal Madd_pulse_spacing_share0000_cy_11_rt_4 : STD_LOGIC;
24.    signal Madd_pulse_spacing_share0000_cy_12_rt_6 : STD_LOGIC;
25.    signal Madd_pulse_spacing_share0000_cy_13_rt_8 : STD_LOGIC;
26.
27.    begin
28.        sync_pulse : FDC
29.            port map (
30.                C => clock20Mhz_BUFGP_364,
31.                CLR => reset_IBUF_647,
32.                D => input_pulse_IBUF_415,
33.                Q => sync_pulse_649
34.            );
35.        pulse_spacing_0 : FDC
36.            port map (
37.                C => clock20Mhz_BUFGP_364,
38.                CLR => reset_IBUF_647,
39.                D => pulse_spacing_mux0003(31),
40.                Q => pulse_spacing(0)
41.            );
```

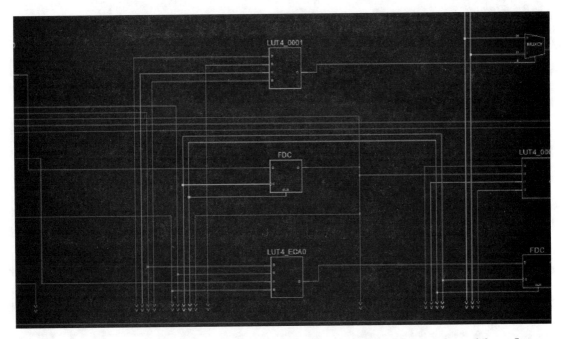

Figure 6–29: Expanded Technology Schematic View (Material based on or adapted from figures and text owned by Xilinx, Inc., courtesy of Xilinx, Inc. Copyright © Xilinx 1995–2008 used in Xilinx ISE WebPack™ software version 10.1.)

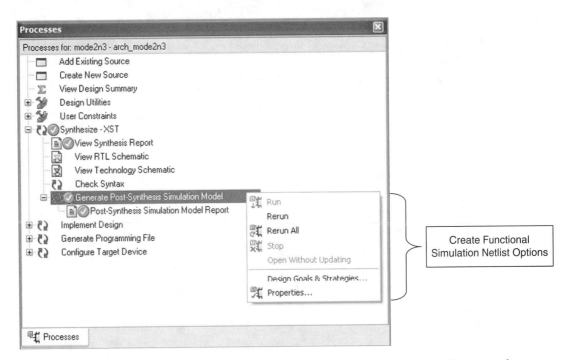

Figure 6–30: Postsynthesis Options (Material based on or adapted from figures and text owned by Xilinx, Inc., courtesy of Xilinx, Inc. Copyright © Xilinx 1995–2008 used in Xilinx ISE WebPack™ software version 10.1.)

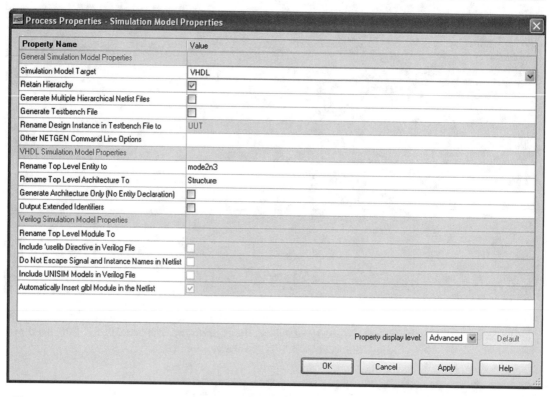

Figure 6–31: Functional Netlist Options (Material based on or adapted from figures and text owned by Xilinx, Inc., courtesy of Xilinx, Inc. Copyright © Xilinx 1995–2008 used in Xilinx ISE WebPack™ software version 10.1.)

6.7. Chapter Overview

The synthesis process takes the high-level design and breaks it down to a mid-level netlist. The design is getting closer to the file that will be used to program the FPGA. The part number and manufacturer must be known, so the synthesis tool can start associating the design with the part's internal resources. Several output files and schematic views provide additional information about the design's current state. Schematic views show what logic connections are necessary to create the design. The postsynthesis simulation netlist contains some predicted timing; real timing is applied during implementation. This file can be used with the original testbench to verify that the synthesis did not change the design. After successfully completing synthesis, implementation is performed.

Key Points to Remember

- Synthesis is required and must be performed prior to implementation.

- Third party synthesis tools output a synthesized netlist for the implementation tool.

- Synthesis automatically feeds into implementation when using a manufacturer's complete package tool.

- Functional simulation should be performed, if time permits.

- The RTL and technology views show what logic makes up the design.

Chapter Link

Xilinx's Webpack ISE 10.1: www.xilinx.com/tools/webpack.htm.

CHAPTER 7

Implementation

7.1. Introduction

Implementation, also referred to as *place and route* (PAR), is the phase in FPGA
development where the design has been synthesized and an RTL simulation performed
(at least I hope), and maybe a functional simulation. The design is no longer at a high level
but is a mid-level netlist format created by the synthesis process. This is the development
process that produces a bit stream file. Implementation can be very time intensive, because so
many elements must be considered, decisions made, and potential issues to resolved.
Some designs are implemented with ease, while others can take days to complete. In my
opinion, the implementation tool has the hardest job of all the development process tools.
So many options and features are available, that can be used to resolve issues or provide a
better placement or design layout. Ultimately, the bit stream file created in this phase is used
to program the FPGA. In the previous phases, interface signals or board layout were not
considered; however, considering these types of things during implementation can make
board layout and interface much easier. By using specific implementation options, signals
can be assisted to specific pins locations to make board layout easier. The synthesized
netlist is the minimum input to the implementation phase; and the output is a bit stream or
programming file, an optional gate-level simulation netlist, and a timing file, see Figure 7–1.

In this chapter, you will learn about

- Implementation process.

- Tools and setup.

- Constraint files.

- How to perform an implementation through a tutorial.

7.2. What Is Implementation?

Implementation is the process that maps the synthesized netlist to the specific or target FPGA's
resources and interconnects them to the FPGA's internal logic and I/O resources. During this

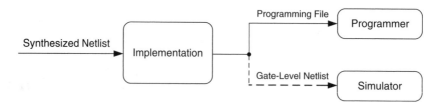

Figure 7–1: Implementation Phase Inputs and Outputs

process, the physical design layout is determined. This is the final development process that manipulates the design before it is programmed into a device. Each manufacturer performs implementation differently, but the concept is basically the same. The process described in this section is like the one performed by Xilinx's implementation tool. The implementation process takes four steps to convert the mid-level netlist to a final programming file—translate, map, place and route, and generate programming file, see Figure 7–2.

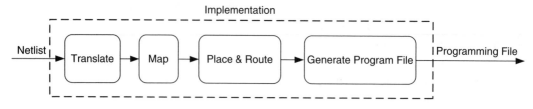

Figure 7–2: Implementation Steps

7.2.1. Translate

The translation process takes the input netlist and merges it with the design constraints (if provided) to create a native generic database (NGD) output file, see Figure 7–3. Acceptable netlist formats depend on the implementation tool, a common one is EDIF, and the manufacturer's specific format (e.g., Xilinx's is NGC). The synthesized input netlist is automatically fed into the translation process when using a complete package development tool. However, the implementation tool must be directed to the location of the synthesized netlist created by a third party synthesis tool. The NGD netlist describes the logical design in terms of Xilinx's primitives. If an error is detected during translation, the tool stops. The error must be corrected and the implementation process must be restarted from the beginning.

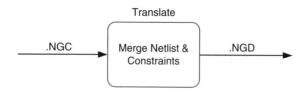

Figure 7–3: Translation Process

Once translation is complete, the NGD output netlist is automatically fed into the mapping process. In addition to the NGD file, an optional post-translation simulation file can be generated. This file is used to verify that the translation process has not changed the design. A post-translation simulation file may not be offered by all implementation tools, so consult your implementation's user's manual. This is an incremental simulation and not the gate-level simulation.

7.2.2. Map

Mapping takes the NGD netlist, which is a logical design, and maps it to the target FPGA, see Figure 7–4. First, a logical design rule check (DRC) is performed on the NGD netlist. Then, the logic is mapped to the target FPGA's logic cells, I/O cells, and other internal resources. Errors encountered during the mapping process cause the implementation tool to stop. All errors must be corrected, and the implementation process must be restarted from the beginning. The output from the mapping process is a native circuit description (NCD) file. NCD is the physical representation of the design that is mapped to the target FPGA's internal resources or components. The NCD is the file that feeds into the place-and-route stage.

In addition to the NCD output file, an optional post-mapping simulation file can be generated. This file is used to verify that the translation process has not changed the design. The post-mapping simulation file may not be offered by all implementation tools, so consult your implementation's user's manual. This is an incremental simulation and not the gate-level simulation.

7.2.3. Place and Route

The place-and-route process takes the NCD file from the mapping process and interconnects the design (places and routes it), see Figure 7–5. After the place-and-route process is complete it outputs an NCD file, which is used to create the bit stream file that is used to program the FPGA. The optional gate-level simulation and timing file can be generated to perform simulation. This gate-level simulation is more meaningful than post-translation and post-mapping simulation, because gate-level simulation files provide actual gate delays based on routing and placement. I would perform the post-translation and post-mapping simulations only for troubleshooting purposes. For example, if a functional simulation was successful but the gate-level one was not, the incremental simulation can help narrow down where the problem first occurred.

Figure 7–4: Mapping Process

Figure 7–5: Place-and-Route Process

7.2.4. Generate Program File

The final implementation step is to generate the programming file. The NCD output file from the place-and-route step is used to create the FPGA's programming file or bit stream, Figure 7–6. It could reside on a nonvolatile device, like a PROM, or within the FPGA device. This file can be automatically downloaded to the FPGA at power-up or when commanded by an external device, like a microprocessor. The process of loading the bit stream into the FPGA is called *configuration*. The datasheet, user's guide, or application notes define the configuration and programming options. Data loads can be a combination of serial, see Figure 7–7, or parallel, with the FPGA acting as the master, see Figure 7–8 (controlling

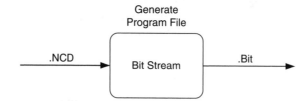

Figure 7–6: Generate Programming File

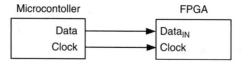

Figure 7–7: Serial Slave Configuration

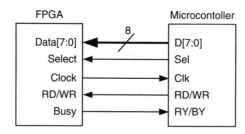

Figure 7–8: Parallel Master Configuration

Figure 7–9: Serial Daisy Chain Configuration

external device), or slave (being controlled by external device). They can be programmed one at a time or daisy-chained to program multiple loads, see Figure 7–9. The FPGA has configuration pins that are used during the configuration process. The manufacturer provides the specific operating details of these pins in its documentation (i.e., datasheet, application note, user's guide). The implementation tool provides various options for creating the bit stream. The bit-stream data can be compressed or uncompressed. Compressed bit-stream data are of a fixed size and the manufacturer provides this information. Oftentimes, security options are available to prevent unauthorized downloading of the bit stream. Once the bit stream is ready, the next step is to program the device that will hold this file.

7.3. Implementation Tools

The tools needed are an implementation tool and a design editor. The implementation tool performs the design implementation or PAR, and the editor is needed for design modification. The implementation tool is offered by the FPGA's manufacturer and generally not a third-party company. These tools use proprietary algorithms to process the synthesized netlist and produce the final programming file. The general setup is pretty simple, even when using a synthesized netlist, from a third-party tool. The synthesized netlist is automatically fed into the implementation process for complete package development tools. On the other hand, the tool must be directed to the synthesized netlist for a third party's netlist. Even though the setup seems easy, working with the tool to get the program file that meets your needs can be challenging. A lot of information about the target FPGA, such as part number, speed, and package, has to be provided to the synthesis tool and is contained in the output synthesized netlist. Because of all the work done by the synthesis process, it seems like implementation should be easy. This is not the case. Putting the design into the FPGA and interconnecting can be the most challenging and time-consuming part of the development process.

7.4. Implementation Inputs

The minimum input to the implementation phase is the synthesized netlist from the synthesizer with an optional user-defined constraint file. When using a third party synthesizer, the netlist must be in a format that is readable by the manufacturer's implementation tool. So, consult your user's manual to make sure the formats are compatible. However, if you are using a complete package development tool, then the tool automatically creates the correct format and feeds it into the implementation process. The user-defined constraint files contain such information as timing, pin assignments, and internal placement for logic.

User-defined constraints put restrictions on the implementation tool and should be used with caution. Constraints make the tool work harder, because it must consider the restrictions that it must follow and still do its job. Implementing a design that utilizes most of the device's resources can greatly increase implementation time and may even cause the process to fail. I am not saying that constraints are bad and should not be used, because they can be necessary. I am just saying to consider all the factors when determining when and what should be constrained. If at all possible, try to keep the device utilization below a reasonable percent. As stated before, high resource utilization increases implementation time and makes it difficult if not impossible for the design to be placed and routed. What is reasonable? I leave it to you to determine a reasonable percent, but consider the room needed for potential growth and spare pins. Some companies predefine how to derate the resources. I like 50%, but that is not always possible, so this number changes from design to design.

Over the years, preassigning pins to an FPGA has been my most used constraint. Assigning pins are most beneficial when they are based on the placement of the components that will interface with the FPGA. The board designer will thank you or you will thank yourself, because this can make board routing so much easier and faster. Many other factors may be considered when determining if the tool or you should assign pins. If you do not care, then I would say first let the tool decide, you always have the option to redefine later. In fact, even if you do care, you can let the tool make the initial pin assignment, review the list, and make changes as necessary. This reduces the restrictions put on the tool, and you still get what you want. Because each situation is different, you should consider the pros and cons and then make your decision.

7.5. Implementation Outputs

Outputs from implementation are the bit stream file and an optional simulation netlist. Implementation also creates a lot of files and directories, some you will care about and others you will not. Some information that you may find important is

- Pin assignments.

- Timing information.

- Number of unrouted signals.

- Errors and warnings.

- Utilized resources.

Xilinx's ISE generates a report file for each of the implementation stages and several other different files within each stage:

- Translate

  Translate report.

  Floor plan design.

  Post-translation simulation model.

- Map

  Mapping report.

  Post-mapping static timing.

  Post-mapping floor plan design.

  Post-mapping simulation model.

- Place and route

  Place-and-route report.

  Clock region report.

  Asynchronous delay report.

  Pad report.

Examples of these files are provided in the tutorial section. For simple designs, you may need only the report that contains the pin assignments. However, for complex designs, where the tool is having difficulty meeting the timing or placing the design, it will be necessary to view some of the other files and use the advanced implementation tool's features.

Now that we have gone through what it takes to get the design through place and route, this is a good time to provide a tutorial.

7.6. Implementation Tutorial

This tutorial continues to use Xilinx's Webpack ISE 10.1 for the FPGA development.

Most if not all complete implementation tools can process the design from synthesis all the way through to generating the programming file without stopping, if no errors are encountered.

Implementation Assumptions

- Preinstalled Xilinx's Webpack ISE 10.1.

- ISE synthesized netlist.

- Basic knowledge of the Webpack ISE.

- Continuous processing from synthesis and Webpack is open.

Expand Implementation Options

Click on the + beside Implement Design to reveal the three implementation steps, see Figure 7–10. Notice that Synthesis – XST has a checkmark, indicating the process was successful.

There are two implementation options: Option 1 performs all three processes in one step, option 2 performs each process individually. Regardless of which option is selected, the same information is provided. If an error occurs during any part of the process, the tool stops, the error(s) must be corrected before continuing, and the entire process must be repeated. Steps for both options are demonstrated in this tutorial. Select one option or try both.

Option 1. Implement All

Select and right click on Implement Design. This shows all processing options, see Figure 7–11.

Select Run or Rerun. This performs entire implementation process (i.e., translation, mapping, and placing and routing) at once.

Note: Selecting Rerun All performs both the synthesis and implementation processes.

Option 2. Individual Implementation

The individual processes are performed one at a time.

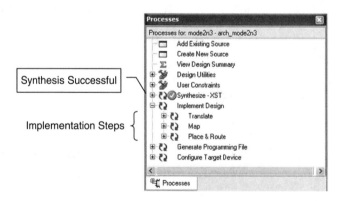

Figure 7–10: Implementation Steps (Material based on or adapted from figures and text owned by Xilinx, Inc., courtesy of Xilinx, Inc. Copyright © Xilinx 1995–2008 used in Xilinx ISE WebPack™ software version 10.1.)

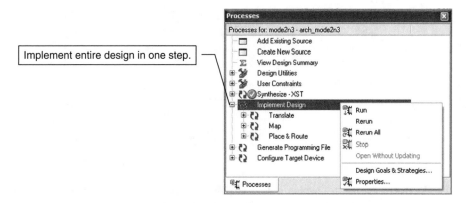

Implement entire design in one step.

Figure 7–11: Implement All (Material based on or adapted from figures and text owned by Xilinx, Inc., courtesy of Xilinx, Inc. Copyright © Xilinx 1995–2008 used in Xilinx ISE WebPack™ software version 10.1.)

Translate Design

Translation is the first of the implementation processes. The input to this process is a synthesized netlist and user-defined constraint file(s). Native generic complier (NGC) is the synthesized netlist format. Since the development tool for this tutorial performs both synthesis and implementation, the netlist is picked up by the tool automatically. At this point, no user-defined constraints have been established. Constraints are not always known or defined in the beginning and can be added later, which is the case for this tutorial. The output is a native generic database or NGD file, which is automatically fed into the mapping process.

Click on the + beside `Translate Design`. This shows that a translate report, floorplan design, and post-translate simulation model file are created and made available after the translate step. Each of these options can be run individually or all at once, as in this case, see Figure 7–12.

Run Translation

Select and right mouse click on `Translate`. Select `Run`.

Translate Report

Once the design has been translated successfully, a checkmark appears beside `Translate`.

Double click on `Translation Report` to see the report in the viewing area, see Figure 7–13. As the design is being translated, the transcript window shows the status, see Figure 7–14.

Floorplan Design

Floor planning is an advanced feature that can allow you to physically locate logic in the FPGA. This feature is not discussed in this book. However, it will be used to for pin assignment.

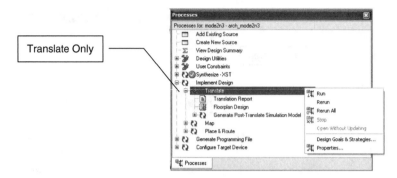

Figure 7–12: Perform Translate Implementation (Material based on or adapted from figures and text owned by Xilinx, Inc., courtesy of Xilinx, Inc. Copyright © Xilinx 1995–2008 used in Xilinx ISE WebPack™ software version 10.1.)

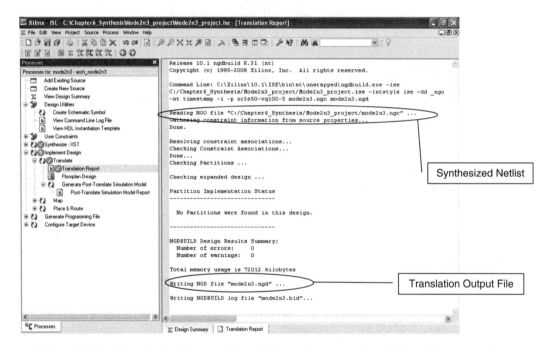

Figure 7–13: Translation Report (Material based on or adapted from figures and text owned by Xilinx, Inc., courtesy of Xilinx, Inc. Copyright © Xilinx 1995–2008 used in Xilinx ISE WebPack™ software version 10.1.)

Generate Post-translate Simulation Model

This produces a simulation netlist or file that allows you to verify that the translation step did not change the design. The post-translate simulation output file may not be an option for all implementation tools, so consult your implementation's user's manual. Run the `Generate Post-Translate Simulation Model` option to create the simulation file. This file is for simulation purposes only and cannot be synthesized.

Figure 7–14: Translation Transcript Status (Material based on or adapted from figures and text owned by Xilinx, Inc., courtesy of Xilinx, Inc. Copyright © Xilinx 1995–2008 used in Xilinx ISE WebPack™ software version 10.1.)

Select and right mouse click on `Generate Post-Translate Simulation Model`. This reveals its options, see Figure 7–15.

Select `Run`.

Once this process is complete, open the report to see the processing information and path to the simulation netlist. For this tutorial, the post-translation simulation is not performed.

Click on the + next to `Generate Post-Translate Simulation Model`.

Double click on `Post-Translate Simulation Model Report`. The report opens in the viewing area, see Figure 7–16.

Now that the translation has successfully been performed, the design will be mapped.

Figure 7–15: Generate Post-Translate Simulation Model (Material based on or adapted from figures and text owned by Xilinx, Inc., courtesy of Xilinx, Inc. Copyright © Xilinx 1995–2008 used in Xilinx ISE WebPack™ software version 10.1.)

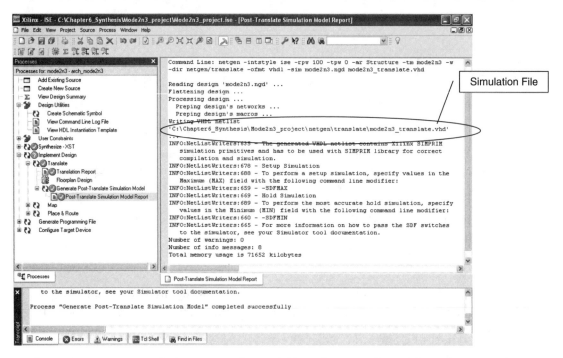

Figure 7-16: Post-translation Report (Material based on or adapted from figures and text owned by Xilinx, Inc., courtesy of Xilinx, Inc. Copyright © Xilinx 1995-2008 used in Xilinx ISE WebPack™ software version 10.1.)

Map Design

The translation output file NGD file is automatically fed into the mapping process. It is used to map the design to the internal FPGA logic.

Select and right mouse click on Map. To view the mapping options, see Figure 7-17.

Select Run.

The mapping process has been successfully completed, which is indicated by the checkmark that appears next to Map in the process window and the Transcript window, see Figure 7-18.

Generate Post-Map Simulation Model

Like the translation process, you have an option to generate a simulation netlist. The purpose of the post-mapping simulation file is to verify that the mapping process has not changed the design. This file is good only for simulation and cannot be used for synthesis. This tutorial shows you how to create the netlist but does not perform this simulation. Personally, I would perform this simulation only for troubleshooting reasons. If problems are found in the gate-level simulation and not in the post-synthesis or functional simulation, then I use this

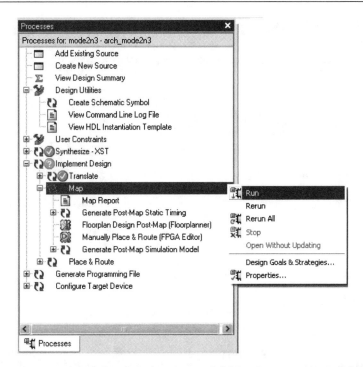

Figure 7–17: Perform Map Implementation (Material based on or adapted from figures and text owned by Xilinx, Inc., courtesy of Xilinx, Inc. Copyright © Xilinx 1995–2008 used in Xilinx ISE WebPack™ software version 10.1.)

or a post-translate simulation file to determine exactly which process changed the design. For now, this is omitted.

Select and right mouse click on `Generate Post-Map Simulation Model`. Select `Run`. The post-mapping simulation netlist is created.

View Post-Mapping Report

Click on the + next to `Generate Post-Map Simulation Model`.

Select and right mouse click on `Post-Map Simulation Model Report` and open file. The report opens in the viewing area, as shown in Figure 7–19, where the path and filename of the simulation netlist is provided.

Place and Route Design

The next implementation step is place and route.

Select and right mouse click on `Place & Route`. This shows the place and route options, see Figure 7–20.

Select `Run`.

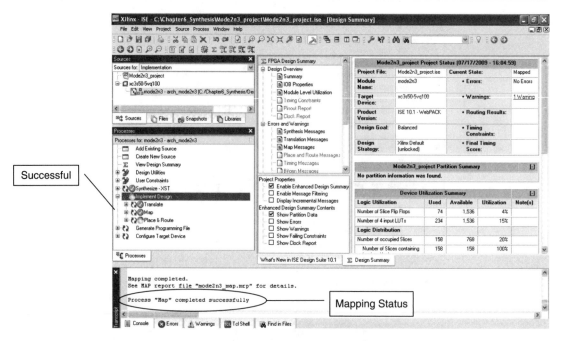

Figure 7–18: Successful Mapping Implementation (Material based on or adapted from figures and text owned by Xilinx, Inc., courtesy of Xilinx, Inc. Copyright © Xilinx 1995–2008 used in Xilinx ISE WebPack™ software version 10.1.)

A checkmark beside `Place & Route` indicate the process was successful, see Figure 7–21.

All the place-and-route reports are now available. There are several reports, but for the most part, unless you are having problems or need to work with other aspects of the tool for constraints or other things, you probably will not even bother opening the other files. The place-and-route and pad reports are generally the ones of interest.

Double click on `Place & Route Report`. This opens the report file in the viewing area, see Figure 7–22. It could have been opened using a standard text editor. The file has the design's name with `.par` extension, so this file is `mode2n3.par`. The report provides details about the place-and-route process, such as device utilization, status of placement, and routing. Place and route could have been performed by typing something like `par -w -intstyle ise -ol std -t 1 mode2n3_map.ncd mode2n3.ncd mode2n3.pcf` on the command line.

The default options were selected for the various levels; however, these options may have to be adjusted for designs that are difficult to place and route. The device utilization is the area of most interest, because it defines how much of the internal resources have been used. Higher utilization makes it more difficult for the implementation tool to work. This summary gives you a good indication whether the part is sufficient or you need to go up a size or two. So, view this area to make sure there is room for growth inside the FPGA.

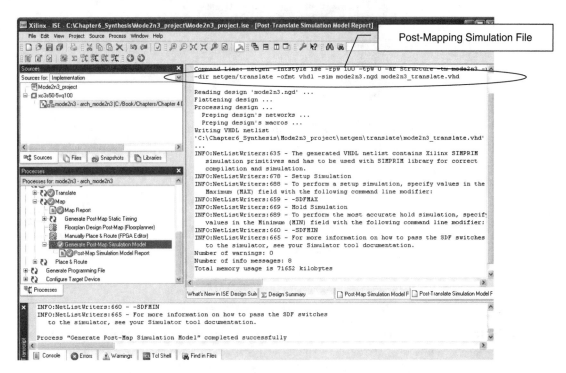

Figure 7–19: Post-map Simulation Netlist (Material based on or adapted from figures and text owned by Xilinx, Inc., courtesy of Xilinx, Inc. Copyright © Xilinx 1995–2008 used in Xilinx ISE WebPack™ software version 10.1.)

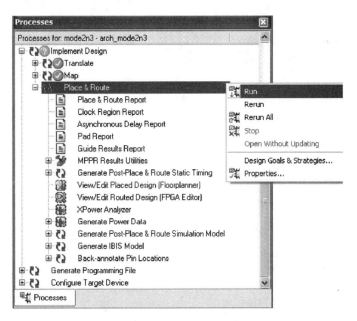

Figure 7–20: Place-and-Route Implementation (Material based on or adapted from figures and text owned by Xilinx, Inc., courtesy of Xilinx, Inc. Copyright © Xilinx 1995–2008 used in Xilinx ISE WebPack™ software version 10.1.)

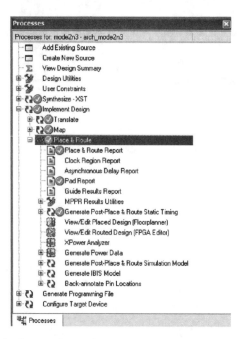

Figure 7–21: Successful Placing and Routing (Material based on or adapted from figures and text owned by Xilinx, Inc., courtesy of Xilinx, Inc. Copyright © Xilinx 1995–2008 used in Xilinx ISE WebPack™ software version 10.1.)

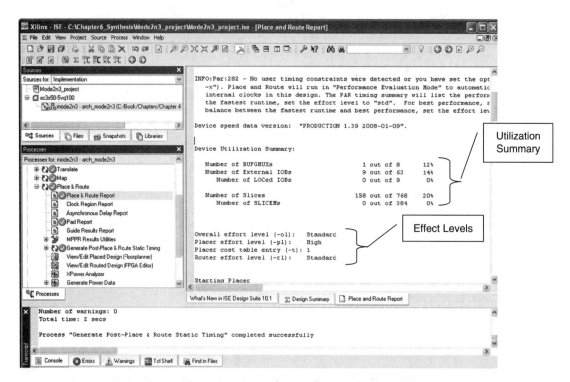

Figure 7–22: Place-and-Route Report (Material based on or adapted from figures and text owned by Xilinx, Inc., courtesy of Xilinx, Inc. Copyright © Xilinx 1995–2008 used in Xilinx ISE WebPack™ software version 10.1.)

View Pin Assignments

Double click on `Pad Report`. The pad reports opens in the viewing area; however, it could have been opened using a text editor, since it is a text file. It contains the filename for the input and output files, part information, and I/O signals information, see Figure 7–23. Each pin on the FPGA package is listed in this file. Information such as pin name, pin number, usage, direction, I/O standard, bank in which it is located, drive current slew rate, and so forth are provided. Initially, the implementation tool was allowed to run without constraints, so the I/O pin assignments were decided by the tool. Table 7–1 shows the I/O pin assignments for the `mode2n3.vhd` design.

Create Constraints

We can make changes in the I/O pin assignments through the constraint file. The tool automatically created a constraint file, since one was not previously provided, Figure 7–24. The implementation tool assigned `reset` to pin 86 and `input_pulse` to pin 92. Let us say the board layout has the interfaces for `reset` and `input_pulse` signals located on the other side of the board. These signals can be reassigned to more appropriate pins, based on the board layout, see Figure 7–25.

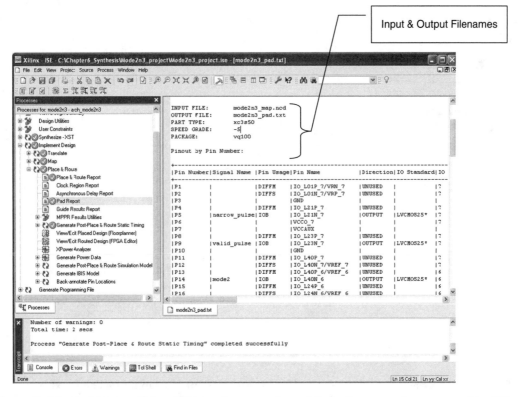

Figure 7–23: Pad Report (Material based on or adapted from figures and text owned by Xilinx, Inc., courtesy of Xilinx, Inc. Copyright © Xilinx 1995–2008 used in Xilinx ISE WebPack™ software version 10.1.)

Table 7–1: Pin Assignments

Pin Name	Pin Number	Direction
narrow_pulse	P5	Output
valid_pulse	P9	Output
mode2	P14	Output
wide_pulse	P85	Output
reset	P86	Input
invalid_mode	P88	Output
clock20Mhz	P89	Input
mode3a	P91	Output
input_pulse	P92	Input

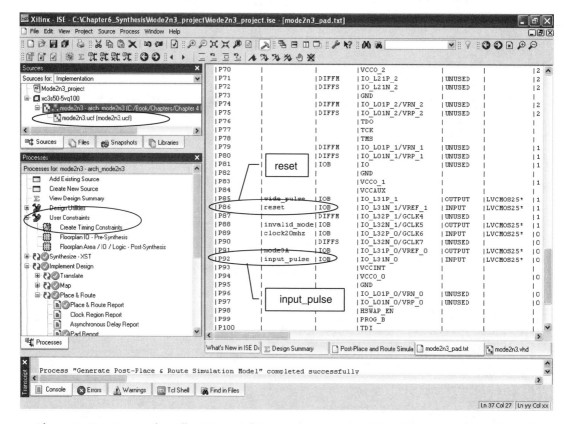

Figure 7–24: Constraint Files (Material based on or adapted from figures and text owned by Xilinx, Inc., courtesy of Xilinx, Inc. Copyright © Xilinx 1995–2008 used in Xilinx ISE WebPack™ software version 10.1.)

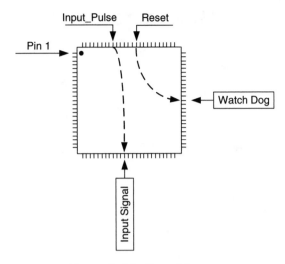

Figure 7–25: Board Layout

Change Reset and Input_pulse Pin Assignments

To change the pin assignments for the reset and input_pulse signals, use the floorplanner. This option is found in the Processes window.

Select and right mouse click on Floorplan Area/IO/Logic-Post-Synthesis in the Processes window. This shows the options, see Figure 7–26.

Select Run. The Floorplan window opens, see Figure 7–27.

Select Package View. This makes the package view active. Holding the cursor over the pins, reveal their types, such as power, ground, clock, etc.

Select Tools → Allow Mode. If in Select Mode, you will not be able to assign pins.

Drag and Drop Pin Assignment

Select the cell block to the left of input_pulse and drag it to P37, see Figure 7–28.

The pin area will become shaded, indicating a signal is located at that pin. Notice that the tool completed the bank information. Additionally, the tool will not allow you to assign a signal to a restricted pin, such as power or ground.

Select and Type Pin Assignment

Select the Loc cell column for reset. This places cursor in a cell for typing, see Figure 7–29.

Type in P63, to indicate the pin location for reset.

Pin 63 is now shaded and no other signal can be placed in this location unless this pin assignment has been removed.

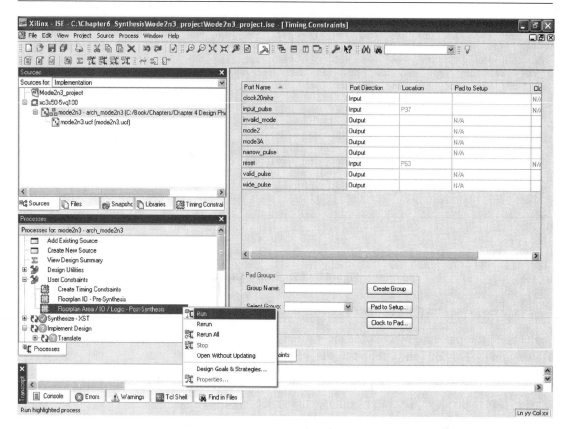

Figure 7–26: Signal Pin Reassignment (Material based on or adapted from figures and text owned by Xilinx, Inc., courtesy of Xilinx, Inc. Copyright © Xilinx 1995–2008 used in Xilinx ISE WebPack™ software version 10.1.)

The constraint file could have been created prior to placing and routing, and sometimes this is necessary. The previous two ways show how easy it is to reassign pins by either drag and drop or selecting and typing. If a user constraint file has not been created and applied to the implementation tool, then the tool automatically creates one, as in this tutorial. Sometimes, it is best to let the tool make the first decision, then reassign as necessary.

Save and Close the Floorplanner

Notice that `Implement Design` in the `Processes` window has a question mark (?), as its status, meaning implementation has to be repeated, see Figure 7–30.

Select and right mouse click on `Implementation Design`. Select `Run`.

After `Implement Design` has been successfully completed, we can view the pin changes in the constraint file.

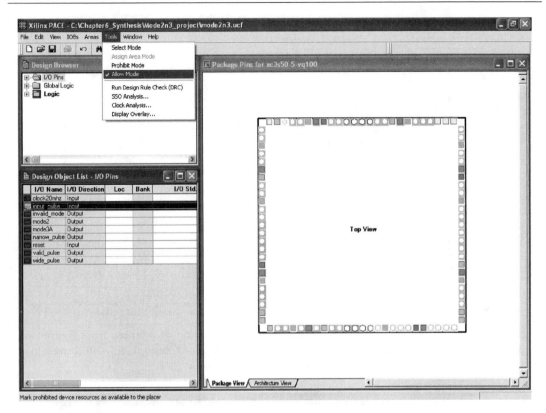

Figure 7–27: Floorplan Window (Material based on or adapted from figures and text owned by Xilinx, Inc., courtesy of Xilinx, Inc. Copyright © Xilinx 1995–2008 used in Xilinx ISE WebPack™ software version 10.1.)

Select and right mouse click on `mode2n3.ucf` on the `Sources` tab in the `Sources` window.

Select Open. The constraint file opens in the viewing area, see Figure 7–31.

Select the `Timing Constraint` tab in the `Sources` window. Since this design has only one constraint file, `Constraint Files` is set to `mode2n3.ucf`. If there were additional constraint files, they would be viewable and selectable by using the pull-down arrow.

Select `Show Constraints` from `Specified File`. This option selection matters only if more than one constraint file is associated with the design.

Select `Port`. The new pin assignments are shown in the viewing area, see Figure 7–32. The pin assignments can be viewed here but not modified.

Notice that the other pins do not show the pin assignments from the pad report. This means that these pins could be reassigned during another place-and-route. This is prevented

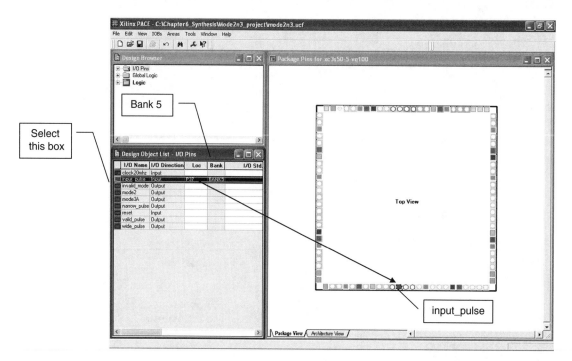

Figure 7–28: Drag and Drop Pin Assignment (Material based on or adapted from figures and text owned by Xilinx, Inc., courtesy of Xilinx, Inc. Copyright © Xilinx 1995–2008 used in Xilinx ISE WebPack™ software version 10.1.)

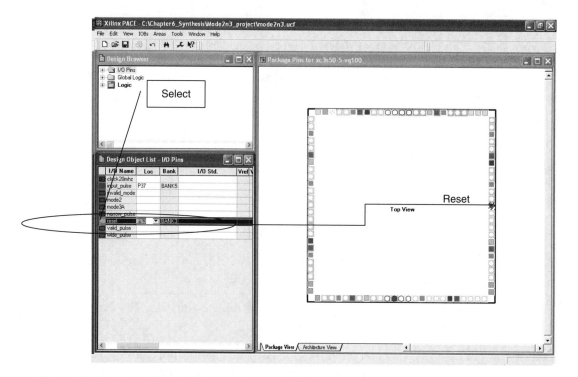

Figure 7–29: Typed Pin Assignment (Material based on or adapted from figures and text owned by Xilinx, Inc., courtesy of Xilinx, Inc. Copyright © Xilinx 1995–2008 used in Xilinx ISE WebPack™ software version 10.1.)

Figure 7–30: Reimplementation (Material based on or adapted from figures and text owned by Xilinx, Inc., courtesy of Xilinx, Inc. Copyright © Xilinx 1995–2008 used in Xilinx ISE WebPack™ software version 10.1.)

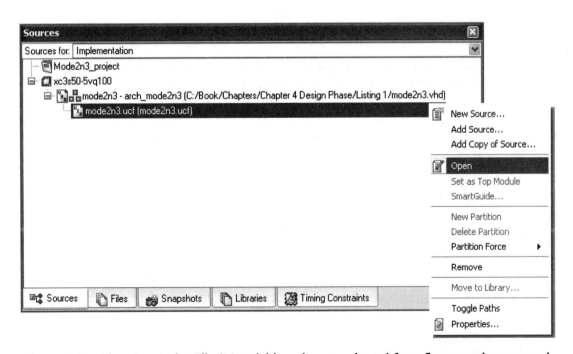

Figure 7–31: View Constraint File (Material based on or adapted from figures and text owned by Xilinx, Inc., courtesy of Xilinx, Inc. Copyright © Xilinx 1995–2008 used in Xilinx ISE WebPack™ software version 10.1.)

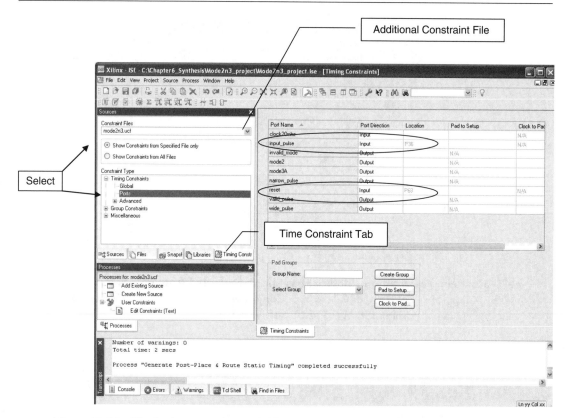

Figure 7–32: Pin Assignment (Material based on or adapted from figures and text owned by Xilinx, Inc., courtesy of Xilinx, Inc. Copyright © Xilinx 1995–2008 used in Xilinx ISE WebPack™ software version 10.1.)

by locking or constraining the pins in a specific location. Using either method, assign (i.e., drag and drop or select and type) the remaining pins to the locations stated in the pad report or some desired location. Then, view the pin assignments in the constraint editor, see Figure 7–33, and the locked pin report, see Figure 7–34.

Pin assignment is only one of many types of constraints that can be added using the floorplanner. The tool can be directed to place specific signals in certain banks. Banks are basically groupings of logic. If you do not want to constrain a signal to a pin, you have the option to constrain it to a bank. An entire book could be written on the different things that can be done during implementation. This chapter has given you a good headstart.

Create Gate-Level Simulation Netlist

Double click on `Post-Place & Route Simulation Model Report`. This will run `Generate Place & Route Simulation Model`, create the gate-level simulation file, and open the report in the viewing area, see Figure 7–35.

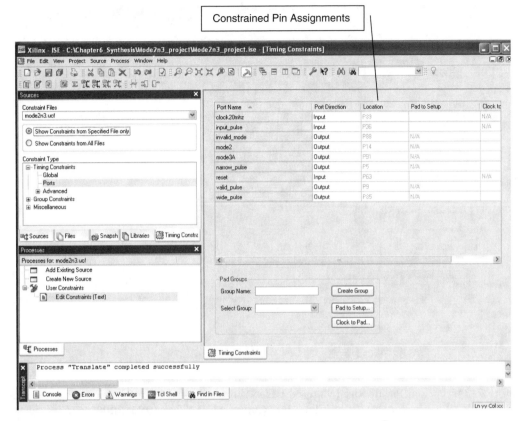

Figure 7–33: Constrained Design Signals (Material based on or adapted from figures and text owned by Xilinx, Inc., courtesy of Xilinx, Inc. Copyright © Xilinx 1995–2008 used in Xilinx ISE WebPack™ software version 10.1.)

Implementation has successfully been completed, as indicated by the checkmark beside `Implement`.

Gate-Level Simulation

Two gate-level files are used for simulation: One is the netlist with file extension `.vhd` and the other contains timing information with file extension `.sdf`. These two files can be used to perform gate-level simulation and verify that the implementation process did not change the design. The same testbench used for RTL and functional simulation can also be used for gate-level simulation. The `sdf` will have to be added to your simulation project, so consult your simulator's user's manual.

Create a Programming File

The final step in implementation is to create the programming or bit-stream file. This file is used to program the FPGA. It may reside in a nonvolatile device or on the FPGA.

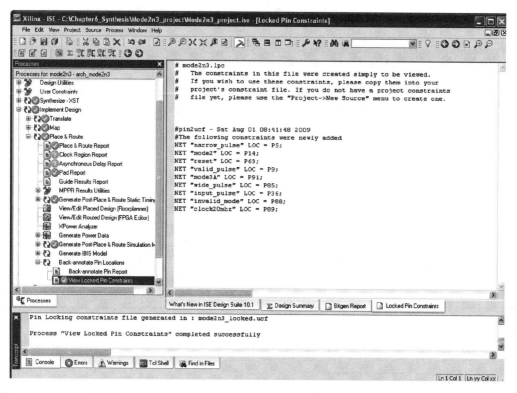

Figure 7–34: Locked Pin Report (Material based on or adapted from figures and text owned by Xilinx, Inc., courtesy of Xilinx, Inc. Copyright © Xilinx 1995–2008 used in Xilinx ISE WebPack™ software version 10.1.)

The datasheet for this device states that an external nonvolatile device should be used to hold the configuration data. Seven modes are supported but slave serial is selected.

Select and right mouse click on `Generate Programming File`. This reveals the several options, see Figure 7–36.

Select `Properties`. In the pop-up window are several categories, each with different options that can be selected, see Figure 7–37. Keep the default selections.

Select `OK`.

Double click on `Programming File Generation Report`. This will create the programming file and open the report in the viewing area, see Figure 7–38. The programming file or bit stream has been successfully created and is named `mode2n3.bit`.

This concludes implementation. The bit-stream file for this design is available and ready to be programmed into a PROM.

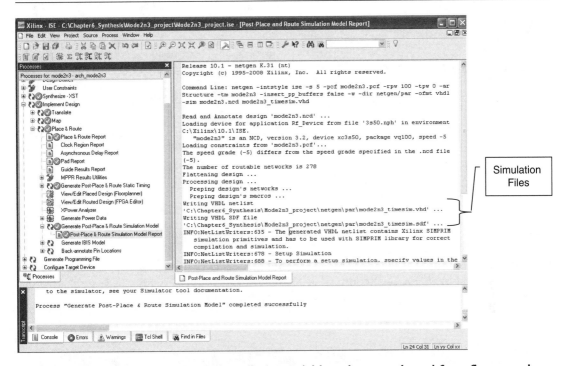

Figure 7–35: Gate-Level Simulation File (Material based on or adapted from figures and text owned by Xilinx, Inc., courtesy of Xilinx, Inc. Copyright © Xilinx 1995–2008 used in Xilinx ISE WebPack™ software version 10.1.)

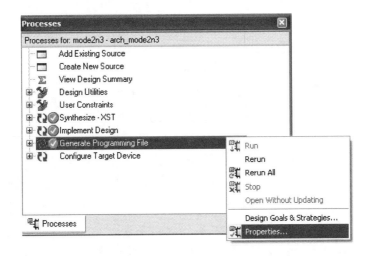

Figure 7–36: Generate Programming File Options (Material based on or adapted from figures and text owned by Xilinx, Inc., courtesy of Xilinx, Inc. Copyright © Xilinx 1995–2008 used in Xilinx ISE WebPack™ software version 10.1.)

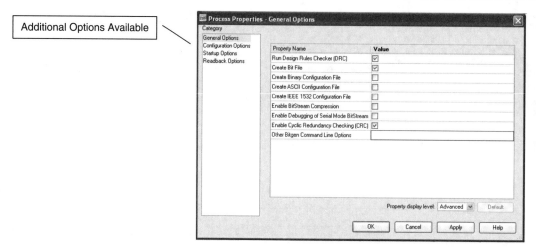

Figure 7–37: Programming File Additional Options (Material based on or adapted from figures and text owned by Xilinx, Inc., courtesy of Xilinx, Inc. Copyright © Xilinx 1995–2008 used in Xilinx ISE WebPack™ software version 10.1.)

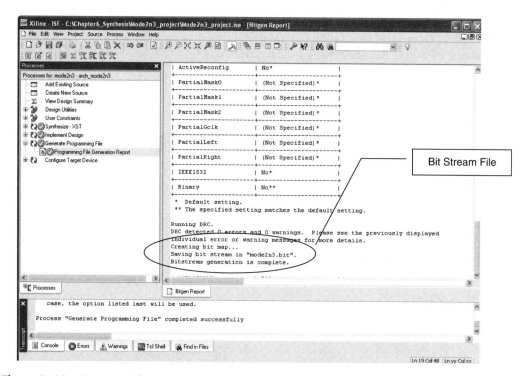

Figure 7–38: Generate Bit Stream (Material based on or adapted from figures and text owned by Xilinx, Inc., courtesy of Xilinx, Inc. Copyright © Xilinx 1995–2008 used in Xilinx ISE WebPack™ software version 10.1.)

7.7. Chapter Overview

Implementation can be a long and complex process. It takes the mid-level netlist and converts it to a file that can be used to program the FPGA. When using a complete package development tool, the synthesized netlist is automatically fed into the implementation phase, while it is necessary to direct the tool to this netlist if it was created by a third party tool. For complex designs, the implementation tools have many options than can be used to overcome place-and-route obstacles. Consult the datasheet, user's guide, or other manufacturer's materials to find the acceptable configuration options for your FPGA.

Implementation Phase Tips

- Remember to lock pin assignments; otherwise they are subject to change.

- Create constraints only when necessary.

- Implementation processes can be performed continuously, if no errors are encountered.

- Use additional tool options to help resolve implementation problems.

Chapter Links

For your convenience, here is a list of links to some free complete development tools:

Xilinx ISE WebPack: www.xilinx.com/tools/designtools.htm.

Altera Quartus II, Web Edition: www.altera.com.

CHAPTER 8

Programming

8.1. Introduction

Programming is the final FPGA development phase and the introduction of hardware. The firmware has been synthesized, simulated (I hope), implemented, and a programming or bit stream file created. This file contains the design's functions and the interconnection information that is used to configure the FPGA. It is now time to take the bit stream and put it into a nonvolatile or volatile memory device. Manufacturers and third party vendors offer programming software, download cables, and programmers that are used to program the specific device. In my opinion, this phase is not as intense; however, it is sometimes very challenging but just as much fun as simulation, because hardware is involved.

In this chapter, you will learn

- Programming options.

- Hardware considerations.

- Programmers options.

- How to program, through a tutorial.

8.2. What Is Programming?

In general, programming involves transferring the bit stream into a nonvolatile or volatile memory device and configuring or programming the FPGA. However, some FPGAs have internal memory and can hold the configuration without an external memory device. Input to the programming phase is a bit-stream or programming file and the output is a programmed device, see Figure 8–1.

Configuration can involve one or a series of daisy-chained or connected FPGAs. Nonvolatile devices, such as PROMs, may be located on the same board as the targeted FPGA (see Figure 8–2) or on another board (see Figure 8–3). The configuration may involve transferring serial or parallel data to the FPGA. The FPGA may be operating in either master (controlling

Doi:10.1016/B978-1-85617-706-1.00008-4

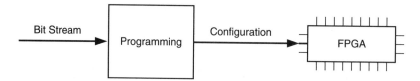

Figure 8-1: Programming Interfaces. Note: Multiple pins are represented by thick pin lines.

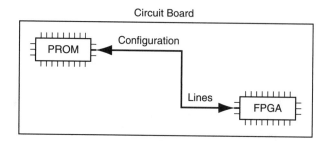

Figure 8-2: Nonvolatile and FPGA on the Same Board. Note: Multiple pins are represented by thick pin lines.

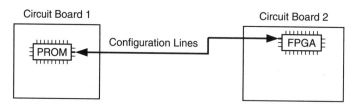

Figure 8-3: Nonvolatile and FPGA on Different Boards. Note: Multiple pins are represented by thick pin lines.

configuration) or slave (not controlling configuration) mode. A configuration guide or similar material provided by the manufacturer specifies the details on the supported configuration and modes. The memory device must have enough memory to hold the design with room for growth. Uncompressed bit-stream size is defined by the manufacturer in the configuration guide or datasheet. For example, Xilinx's configuration guide states that the XC3S50A/AN uncompressed size is 437,312 configuration bits. A two-step process is used to configure the FPGA with the bit-stream file when using an external memory device. In step 1, the bit-stream file is transferred to the memory device; and in step 2, the memory device configures the FPGA.

A word to the wise, for parallel data transfers, make sure you know which bit is transmitted first (i.e., MSB or LSB). Once, I was working with a group that assumed the wrong bit order. The error occurred because the designer assumed the bit order was consistent with his past experience. Unfortunately, this was not the situation, and the bit order was reversed. It took a lot of hours of troubleshooting the board, reading many application notes and other material, before discovering the error. The worst part was that the board layout would not work for this assumption. So the only corrective action was to have someone write a software program to reverse the bit order.

Each time a change was made, the bit order had to be reversed prior to reprogramming, at least until the next revision of the board. In this case, the assumption cost money and time.

8.3. Tools and Hardware

The tools needed for programming depend on the selected memory device. If a microprocessor holds the bit-stream file, then it is merged with the software build. The processor configures the FPGA on power-up or at a specific time. For nonvolatile memory devices, such as PROMs, programming options include a joint test advisory group (JTAG), in-system programming (ISP), and third-party programmers.

8.3.1. Joint Test Advisory Group

IEEE 1149.1, Standard Test Access Port and Boundary Scan Architecture, commonly referred to as *JTAG*, is access pins or ports on a JTAG-compatible device that provide visibility inside the device. A lot of times the terms *JTAG* and *boundary scan* are used interchangeably. Tools needed for JTAG are JTAG software and a software host, and the hardware is a JTAG cable. The host is where the software is located, like a personal computer. The JTAG cable may be a USB connection to the computer with a JTAG connector on the other end. JTAG software is the interface used to transfer the bit stream from the host to the programmable device. The programmable device is connected to a JTAG mating connector, where the JTAG connector is connected. These tools are available from the manufacturer or a third party vendor.

As FGPA packages move away from leaded through-hole parts to surface mount packages (i.e., leadless), it becomes more difficult to use standard manufacturing testing equipment, like the bed of nails to verify populated boards. This was especially difficult on boards with ball grid arrays (BGAs); and this greatly reduced the testability, which lowered the percent of testable area coverage. Therefore, JTAG was developed as a testing and debugging mechanism. It is used to detect manufacturing faults on populated boards. However, over time, it was realized that JTAG ports could be used for programming. Many devices support JTAG, which is indicated on the FPGA's datasheet. The pins are defined and labeled as

- TDI (test data in).

- TDO (test data out).

- TCK (test clock).

- TMS (test mode select).

- Optional TRST (test reset).

Oftentimes, these pins can be used for I/O after configuration. Personally, I never liked reusing them, but you may have a different opinion. A JTAG programming scenario involves transferring the bit stream from a host through the JTAG cable to a header, test pins, or a

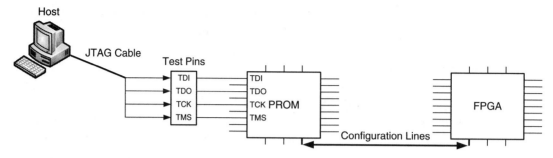

Figure 8–4: JTAG Setup. Note: Multiple pins are represented by thick pin lines.

connector on a board that connects to the JTAG-compatible nonvolatile memory device, see Figure 8–4. The JTAG device configures the slave mode FPGA on power-up. JTAG also supports daisy-chain configurations. Generally, a GUI is provided to guide you through the program process. FPGA manufacturers generally offer JTAG programming tools, cables, and any necessary supplies. Often, they also provide a list of third party vendors or distributors where you can purchase JTAG materials. JTAG can be a great way to quickly incorporate design changes. I found this way to work well with demo boards, prototypes, and breadboards.

8.3.2. In-System Programming

In ISP, the device does not have to be removed from the system or board to be programmed. Sometimes, the device can be programmed while the system is still operating. The datasheet specifies whether the device supports ISP. The tool needed for ISP is software on a host, and the hardware is a download cable. Programming can be done by connecting test pins to automated test equipment (ATE) or a board connector. Some of the supported protocols are the IEEE Standard for Boundary-Scan-Based In-System Configuration of Programmable Devices (IEEE 1532), JTAG, and a serial peripheral interface (SPI).

An ISP programming scenario follows like this: Use the download cable from the host to connect to a board's test pins, header, or connector; then download the bit stream in either the ISP-capable nonvolatile memory device or the FPGA, see Figure 8–5. If the FPGA has volatile memory it must be reprogrammed whenever power is cycled (on → off → on). There are many good reasons to use this option if available. If the FPGA is located in an area that is hard to reach or the device is difficult to remove but access to its programming connector is easy, then ISP is the best programming option. Then, you need only to connect to the easy-access connector for ISP programming.

8.3.3. Third Party Programmers

There are a variety of hardware and software programming options available from third-party manufacturers. They may consist of a GUI software package that can be loaded onto a host, a programming base that connects to the computer, and some socket adaptors or an all-in-one

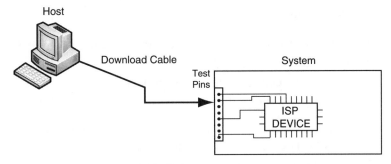

Figure 8–5: ISP Setup

programmer. The programming base is what holds the socket adaptor for programming. A socket adaptor is where the programmable device is placed to get programmed. Each socket adaptor is designed to hold a specific package style. For example, you cannot put a 676 fine-pitch ball grid array (FBGA) into the socket for a 100-pin thin quad flat pack (TQFP).

Companies like Data I/O offer manual and automated programmers. Manual programmers require the user to manually program (hit the `Program` button) for each batch of devices. It does not automatically program one batch after the other. Manual programmers are ideal for low-volume programming. If you have one or just a few devices, then a manual programmer should be sufficient. The programmer can program one or more devices at a time, depending on the size of the adaptor. Programmers come with sockets to fit specific packages; and other ones can be purchased if your package is not supported. Manual programmers come as all-in-one or complete programming packages, while others require you to purchase software that loads onto a computer or host, a programming base, base-computer cable, and sockets. This programming option is best used for programmable devices that are in sockets (generally, boards like prototypes and breadboards), because typically sockets are not allowed on production boards.

Automated programmers are best for production or medium- to high-volume programming. They can program faster and more devices than manually. Automated programmers vary in size and supported packages, and some are designed to easily integrate into production systems. Production programmers are designed for an automated environment, meaning they generally have advanced features not found on the ones best for low volume. For example, they may have the ability to perform pick-and-place functions or simultaneously program multiple devices using different bit-sream files.

Manufacturers offer programming tools for their devices, and third party companies offer a wider variety of programmers and other supplies that support many different manufacturers.

Additional information about Data I/O can be found at www.data-io.com/index.asp.

8.4. Hardware Configuration

As a part of programming, the FPGA can act as either the master or the slave for configuration. The configuration pins on the FPGA are set to specific values (high or low) to indicate whether it is the master or the slave. The configuration guide or datasheet defines the configuration pin settings. For example, Table 8–1 shows the configuration pins' settings for Spartan 3 modes.

It is always a good idea to make programming pins accessible via test points, test pads, or connectors, see Figure 8–6. This can make troubleshooting and programming easier. Even though some FPGA manufacturers offer internal logic analyzer features (not discussed in this book), it is good to keep troubleshooting and debugging in mind when designing a board.

When BGAs were fairly new in FPGA packaging, it became common practice for my digital codesigners to sacrifice the area under the FPGA to expose all the balls. At this time, internal debugging tools were new and not many companies offered them, this made troubleshooting very difficult. I learned a lot about the dos and don'ts when it came to board design and troubleshooting consideration.

Since this is my first real opportunity to talk about hardware and board design, I am going to give you some good general tips I learned along the way. Although board design is not a

Table 8–1: Configuration Pins

Mode	M2	M1	M0
Master serial	0	0	0
Slave serial	1	1	1
Master parallel	0	1	1
Serial parallel	1	1	0
JTAG	1	0	1

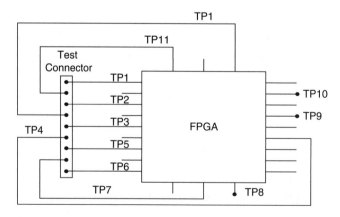

Figure 8–6: Test Pads and Connector for Troubleshooting

development stage, it is very important and really cannot be left out of the process, especially when you consider that the FPGA is hardware and most likely will be verified in a lab and need some troubleshooting.

Tip 1. When daisy-chaining devices, make sure to add the ability to jump out or remove any of the devices, if necessary. Sometimes, a device may cause problems, and having the ability to remove it from the chain helps in isolating the problem.

Tip 2. It is always a good idea to design in some debugging and troubleshooting mechanisms, such as test points, pads, or connectors. Most likely, a logic analyzer will become your best friend for verifying, troubleshooting, or just working with the design. Consider using test connectors that mate with lab equipment hardware. This makes your life so much easier. One of my favorites is a Mictor connector that mates directly with Tektronix's logic analyzer. This made life so much easier than having flying leads soldered to the board. Plus, with surface mount packages, probing is very difficult and, in some cases, impossible. The test connector can always be removed from production boards. As shown in the simulation phase, lab data can be read into a testbench using a connector, like a Mictor, which makes capturing the data a lot easier.

Tip 3. Many FPGAs require several different voltages (such as supply or signal), so be sure to consult the datasheet for the acceptable ranges and the appropriate supply capacitors.

Tip 4. Sometimes, it is desirable to select a specific FPGA package based on the ability to upgrade to a larger size (more internal resources) in the same package without respinning or re-laying out the board. If this is the case, then make sure the two devices are pin-for-pin compatible. The power and grounds may be in different locations. So, if the plan is to start with one specific FPGA size with the goal of being able to replace it with a larger size without having to redo the board, at a minimum, make sure both selections are pin compatible and have the same voltage requirements.

Tip 5. If you have unused input pins, make sure to read the user's guide, application notes, or other appropriate material to determine if it is necessary to connect them to a known state. This can be as simple as making a selection in the implementation tool. Although this is simple, it does require some action on your part. Once, I was asked to help troubleshoot a programmable logic device design created by a subcontractor that was causing a lot of random problems and delaying other critical project activities. I discovered that the designer had not properly terminated the unused inputs. All that was needed to fix this problem was selecting an option in the implementation tool that tied the unused inputs to a known state. As simple as it seems, if you do not know what to look for, this could stop you from getting a design working properly.

The bit-stream or programming file created for the `mode2n3` design is ready to be transferred into a nonvolatile device or the FPGA. Now let us get ready to download it to hardware.

8.5. Programming Tutorial

In this tutorial, Xilinx's Webpack ISE 10.1 is used to transfer the `mode2n3` bit-stream file directly into the FPGA via JTAG programming. The FPGA configuration and lab verification are not covered by this tutorial.

Programming Assumptions

•   Bit stream file available.

•   FPGA is connected to the JTAG connector.

•   JTAG cable.

•   JTAG cable is connected to PC from demo board.

•   Basic knowledge of the Webpack ISE.

Configure Device

Select and right mouse click on `Configure Target Device`. This shows the options for the configure target device process, see Figure 8–7.

Select `Run`. Because no project file has been established, the pop-up window in Figure 8–8 appears. This will allow you to create a project.

Select `OK`.

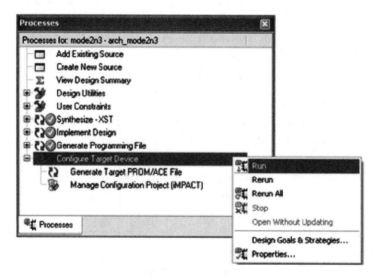

Figure 8–7: Configure Device (Material based on or adapted from figures and text owned by Xilinx, Inc., courtesy of Xilinx, Inc. Copyright © Xilinx 1995–2008 used in Xilinx ISE WebPack™ software version 10.1.)

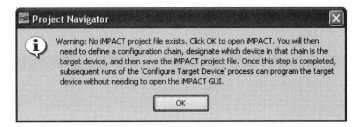

Figure 8-8: Create Configuration Project (Material based on or adapted from figures and text owned by Xilinx, Inc., courtesy of Xilinx, Inc. Copyright © Xilinx 1995-2008 used in Xilinx ISE WebPack™ software version 10.1.)

Make sure `Configure devices using Boundary Scan (JTAG)` is selected, see Figure 8-9.

Select `Finished`.

Add Programmable Device to JTAG Chain

Right click in the open area to add the programmable device. This shows the different JTAG options available, see Figure 8-10.

Select `Add Xilinx Device`.

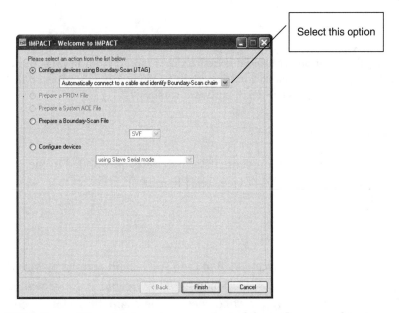

Figure 8-9: JTAG Setup (Material based on or adapted from figures and text owned by Xilinx, Inc., courtesy of Xilinx, Inc. Copyright © Xilinx 1995-2008 used in Xilinx ISE WebPack™ software version 10.1.)

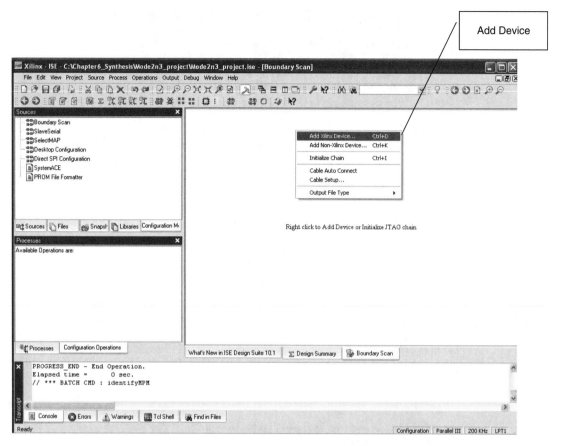

Figure 8–10: JTAG Options (Material based on or adapted from figures and text owned by Xilinx, Inc., courtesy of Xilinx, Inc. Copyright © Xilinx 1995–2008 used in Xilinx ISE WebPack™ software version 10.1.)

Adding Bit-Stream File

The bit-stream file created by the implementation phase is now added to the project so it can be transferred to the programmable device, see Figure 8–11.

Select `mode2n3.bit`.

Select `Open`.

The window shows the FPGA device added to the JTAG chain and the bit stream that will be used for programming, see Figure 8–12. Since the JTAG cable is connected, the device is automatically detected. If there were more devices in the chain, they would be shown as well.

Right click *or* select the symbol of the Xilinx device. The `Processes` window shows your programming options, see Figure 8–13.

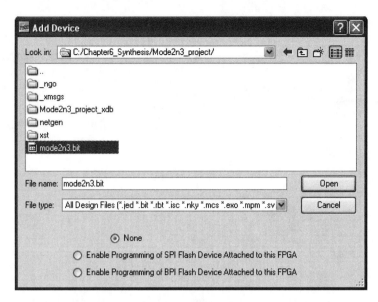

Figure 8–11: Bit Stream Added to Project (Material based on or adapted from figures and text owned by Xilinx, Inc., courtesy of Xilinx, Inc. Copyright © Xilinx 1995–2008 used in Xilinx ISE WebPack™ software version 10.1.)

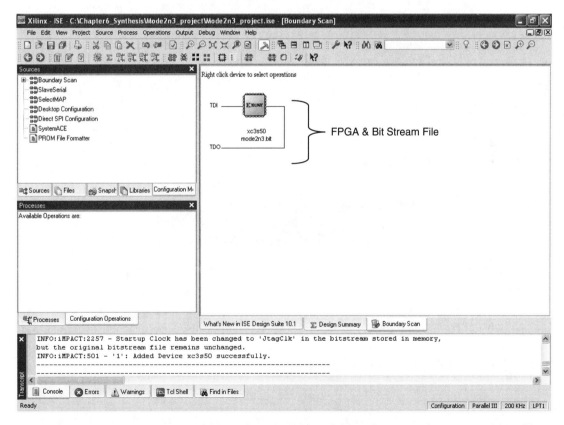

Figure 8–12: JTAG Chain (Material based on or adapted from figures and text owned by Xilinx, Inc., courtesy of Xilinx, Inc. Copyright © Xilinx 1995–2008 used in Xilinx ISE WebPack™ software version 10.1.)

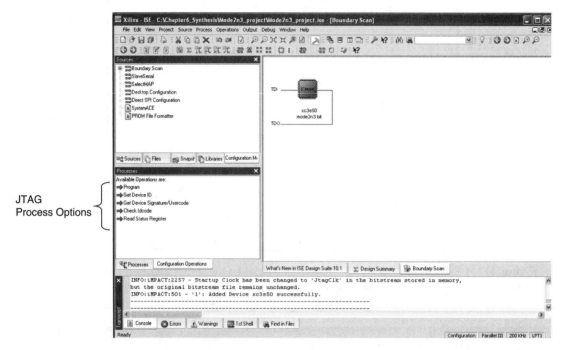

JTAG
Process Options

Figure 8–13: Programming Options (Material based on or adapted from figures and text owned by Xilinx, Inc., courtesy of Xilinx, Inc. Copyright © Xilinx 1995–2008 used in Xilinx ISE WebPack™ software version 10.1.)

Right click on the FPGA icon.

Select Program.

Once the device has been successfully programmed, a successful programming message will appear:

Program Succeeded

Now the FPGA is programmed and ready to be tested. You can use a logic analyzer to verify the design. An automatic testbench can be written to read the lab data, compare them with the simulated data, and write the results to a file or computer monitor. Many options are available for programming an FPGA, the JTAG method used in this tutorial is just one.

8.6. Chapter Overview

The programming phase is exciting because you are starting to work with hardware. All the hard work put into developing simulating, synthesizing, and implementing the design brings you to the point where the design can be viewed in a real-world lab environment.

Good stuff. It is nice to see the simulation, but everything really comes together when you are working in lab.

One book cannot cover all the things that can be done or go wrong while developing an FPGA. However, this book provides some of them, with a good foundation to help you get started and information on where to look for help or guidance. So, if your next step is to create a new design, modify, troubleshoot, or just understand it, you are now prepared to overcome whatever you may encounter. Over time and with each design experience, your skills and knowledge will increase, because each design has the potential to present you with unique and sometimes exciting challenges.

Key Programming Phase Tips

- Add test connectors, pins, and pads to the board to help with verification and troubleshooting.

- For parallel configuration, do not assume the order of data bit transfer (MSB, LSB).

- Select a memory device with sufficient room for growth.

- Make good use of configuration guides and other materials provided by the manufacturer.

Chapter Links

Data I/O: www.data-io.com/index.asp.

References and Sources

Synplify with HDL Analyst User Guide, Release 5.1, with HDL Analyst ®, VHDL, and Verilog Synthesis for FPGAs and CPLDs, available from

Synplicity, Inc.

610 Caribbean Drive

Sunnyvale, California 94089

(408) 548–6000

(408) 548–0050 fax

www.synplicity.com

Essential VHDL: RTK\L Synthesis Done Right, by Sundar Rajan. Copyright © 1997, by Sundar Rajan and Gennis Piazza. All rights reserved. Printed in the USA.

Secondary Surveillance Radar, by Michael C. Stevens, chief scientist, Aviation Systems Division, Cossor Electronics Ltd. (Boston and London: Artech House).

Web Sites

- Xilinx (www.xilinx.com). Xilinx is a well-known leader in FPGA technology. Its Web site provides tons of information that contributed to almost every chapter in the book. You will find information such as FPGA datasheets, application notes, development tools, demo board, and so much more at the site. You can also download the Xilinx free complete development tool, which includes a Xilinx version of the Mentor Graphic's ModelSim simulation tool. Because of the Internet information and free development tool, the tutorials in Chapters 5, 6, and 7 were made possible. I highly recommend downloading the free tool. It is a great way to get started. You will find a wealth of information on this site. It is a great place to look when considering your FPGA options.

- Altera (www.altera.com). Altera's MaxplusII was my first experience using FPGA tools, and its AHDL was my first introduction to HDL. So it is only natural for me to include Altera in my book. While Quartus II has basically replaced Maxplus II, downloading the Web version and playing around with its features was a lot of fun. Chapter 5 shows some simulation screen shots from the Quartus II development tool. In my opinion, this tool is as easy to use as MaxPlus II. I suggest downloading the free version and giving it a try. You will find some of the same standard information such as datasheets, applications, demo boards, and other product information on the site. This is also a great place to look when considering your FPGA options.

- Tektronix and the Moving Pixel Company (www.tek.com and www.movingpixel.com/main.pl?home.html). Tektronix and the Moving Pixel Company Web sites provided information on my favorite logic analyzer and the PC–logic analyzer interface. These sites were most helpful for getting the information and software downloads used in Chapter 5. This information was used to show how easy it is to use data taken with a logic analyzer and import it into your PC for design troubleshooting or verification. I think the PC interface is fun to play with. Go to the site and download the tool. I think you will enjoy it, too.

- Mentor Graphics (www.model.com). Mentor Graphics offers various tools suited for almost any design. The site provided the information used to show the different development tool features discussed in Chapter 6.

- Synopsys (www.synplicity.com). Synopsys makes Synplify, probably the most popular synthesis tool. It offers a variety of product information on the Web site. You will find some of this information about Synopsys's tools in Chapter 6.

- HDL Works (www.hdlworks.com). HDL Works Scriptum located at www .translogiccorp.com/index.html. The HDL Works site offers a lot of products. After comparison shopping, I decided to go with its free HDL editor. It offered a lot of features I like and need as I develop code and the price was good. The primer in Chapter 1 discusses some of the information found on HDL's text editor. This is another great source to get a good tool to help you in developing FPGAs.

Data I/O

- Data I/O (www.data-io.com/index.asp). Data I/O offers a wide range of programming equipment. On its site, you will find more than enough information to help you select the programmer that is best for your application. This site helps provide information on the programming options discussed in Chapter 8.

- Doulos's code generator (www.doulos.com/knowhow/perl/testbench_creation) and Symphony EDA (www.symphonyeda.com). Doulos's Web site offers free code generator tools in addition to other products. However, I thought that, since automatic code generators are becoming more popular, it is worth discussing. So, in Chapter 4, you will find information about Doulos's auto code generator.

Testbenches

A–1. Adder and Subtractor Testbench

Use this testbench to verify the adder and subtractor design in Chapter 2. An example of the simulated output is shown in Figure A–1.

Run the simulation for 200.00 nsec.

```
Library IEEE;
Use IEEE.std_logic_1164.All;
Use IEEE.std_logic_unsigned.All;

Entity testbench Is End testbench;

Architecture tb_MathematicalOperators Of testbench Is

Signal number_1        : std_logic_vector(3 Downto 0) := "0100";
-- setting initial value
Signal number_2        : std_logic_vector(3 Downto 0) := "0010";
-- setting initial value
Signal sum             : std_logic_vector(3 Downto 0);
Signal difference      : std_logic_vector(3 Downto 0);

Component MathematicalOperators Port (
  number_1      : In std_logic_vector(3 Downto 0);
  number_2      : In std_logic_vector(3 Downto 0);
  sum           : Out std_logic_vector (3 Downto 0);
  difference    : Out std_logic_vector (3 Downto 0));
End Component;
```

Begin

mathfunctions: MathematicalOperators
Port Map (
  number_1   => number_1,
  number_2   => number_2,
  sum   => sum,
  difference   => difference);

-- assigning values to the input signals
number_1   <=   "0010" **After** 50.00 nsec,
                         "0111" **After** 100.00 nsec,
                         "0110" **After** 150.00 nsec;

number_2   <=   "0101" **After** 75.00 nsec,
                         "0001" **After** 125.00 nsec;

END tb_MathematicalOperators;

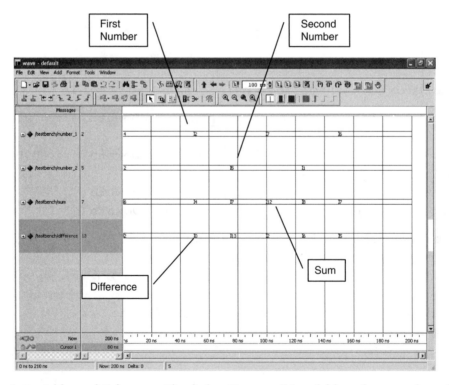

Figure A–1: Adder and Subtractor Simulation Outputs (Material based on or adapted from figures and text owned by Xilinx, Inc., courtesy of Xilinx, Inc. Copyright © Xilinx 1995–2008 used in Xilinx ISE WebPack™ software version 10.1.)

A–2. Logic Gates Testbench

Use this testbench to verify the logic gates design in Chapter 2. An example of the simulated output is shown in Figure A–2.

Run simulation for 300.00 nsec.

```
Library IEEE;
Use   IEEE.std_logic_1164.All;
Use   IEEE.std_logic_unsigned.All;

Entity testbench IS END testbench;

Architecture tb_LogicGates Of testbench Is

Signal number_1      : std_logic_vector (3 Downto 0) :="0001";
Signal number_2      : std_logic_vector (3 Downto 0) :="0101";
Signal or_out        : std_logic_vector (3 Downto 0);
Signal nor_out       : std_logic_vector (3 Downto 0);
Signal and_out       : std_logic_vector (3 Downto 0);
Signal nand_out      : std_logic_vector (3 Downto 0);
```

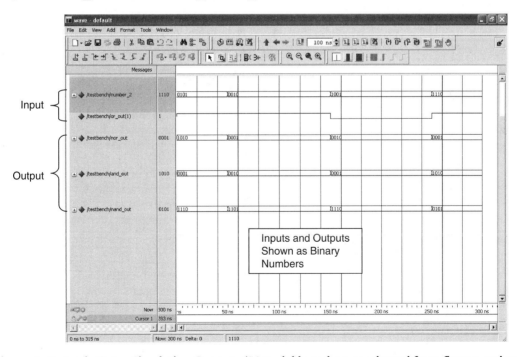

Figure A–2: Logic Gates Simulation Outputs (Material based on or adapted from figures and text owned by Xilinx, Inc., courtesy of Xilinx, Inc. Copyright © Xilinx 1995–2008 used in Xilinx ISE WebPack™ software version 10.1.)

```
Component LogicGates Port (
  number_1   : In std_logic_vector (3 Downto 0);   -- setting
  initial value
  number_2   : In std_logic_vector (3 Downto 0)   -- setting
  initial value
  or_out     : Out std_logic_vector (3 Downto 0);
  nor_out    : Out std_logic_vector (3 Downto 0);
  and_out    : Out std_logic_vector (3 Downto 0);
  nand_out   : Out std_logic_vector (3 Downto 0));
End Component LogicGates;

Begin

LogicGates1: LogicGates
Port Map (
  number_1       => number_1,
  number_2       => number_2,
  or_out         => or_out,
  nor_out        => nor_out,
  and_out        => and_out,
  nand_out       => nand_out);

- assigning values to the input signals
number_1          <=   "1110" After 50.00 nsec,
                       "0101" After 150.00 nsec,
                       "1010" After 250.00 nsec;

number_2          <=   "0010" After 50.00 nsec,
                       "1001" After 150.00 nsec,
                       "1110" After 250.00 nsec;

END tb_LogicGates;
```

A–3. D Flip-Flop Testbench

Use this testbench to verify the D flip-flop design in Chapter 2. An example of the simulated output is shown in Figure A–3.

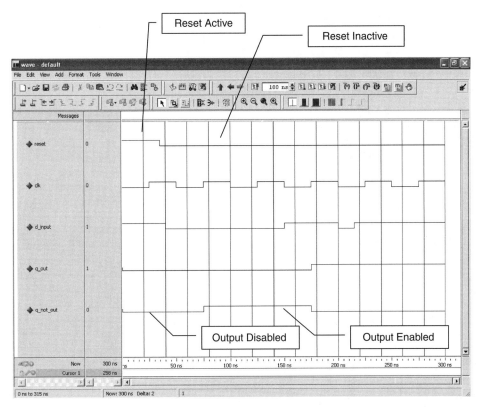

Figure A–3: D Flip-Flop Simulation Outputs (Material based on or adapted from figures and text owned by Xilinx, Inc., courtesy of Xilinx, Inc. Copyright © Xilinx 1995–2008 used in Xilinx ISE WebPack™ software version 10.1.)

Run the simulation for 300.00 nsec.

```
Library IEEE;
Use IEEE.std_logic_1164.All;
Use IEEE.std_logic_unsigned.All;

Entity testbench Is End testbench;

Architecture tb_Dff Of testbench Is

Signal reset      : std_logic := '1';  -- setting initial state value
Signal clk        : std_logic := '0';  -- setting initial state value
Signal d_input    : std_logic := '1';  -- setting initial state value
Signal q_out      : std_logic;
Signal q_not_out  : std_logic;

Constant twenty_five_nsec    : time := 25 nsec;
```

```
Component Dff Port (
  reset          : In std_logic;
  clk            : In std_logic;
  d_input        : In std_logic;
  q_out          : Out std_logic;
  q_not_out      : Out std_logic);
End Component Dff;

Begin

dff1: dff
Port Map (
  reset          => reset,
  clk            => clk,
  d_input        => d_input,
  q_out          => q_out,
  q_not_out      => q_not_out);

create_twenty_Mhz: Process
Begin
Wait For twenty_five_nsec;
 clk <= NOT clk;
End Process;

-- assigning values to input signals
reset           <=     '0' After 35.00 nsec;

d_input         <=     '0' After 40.00 nsec,
                       '1' After 150.00 nsec,
                       '0' After 200.00 nsec,
                       '1' After 215.00 nsec;
END tb_dff;
```

A–4. DFF with Synchronous Enable Testbench

Use this testbench to verify the D flip-flop with synchronous enable design in Chapter 2. An example of the simulated output is shown in Figure A–4.

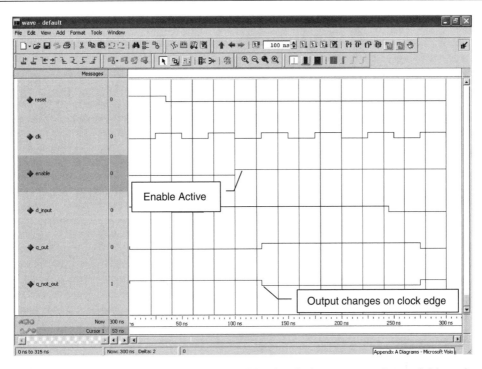

Figure A–4: D Flip-Flop with Synchronous Enable Simulation Outputs (Material based on or adapted from figures and text owned by Xilinx, Inc., courtesy of Xilinx, Inc. Copyright © Xilinx 1995–2008 used in Xilinx ISE WebPack™ software version 10.1.)

Run the simulation for 300.00 nsec.

```
Library IEEE;
Use IEEE.std_logic_1164.All;
Use IEEE.std_logic_unsigned.All;

Entity testbench Is End testbench;

Architecture tb_DffSynEa Of testbench Is

Signal reset     : std_logic := '1';      -- setting initial value
Signal clk       : std_logic := '0';      -- setting initial value
Signal enable    : std_logic:= '0';       -- setting initial value
Signal d_input   : std_logic := '1';      -- setting initial value
Signal q_out     : std_logic;
Signal q_not_out : std_logic;

Constant twenty_five_nsec: time := 25 nsec;

Component DffSynEa Port (
  reset          : In std_logic;
  clk            : In std_logic;
```

```
  enable           : In std_logic;
  d_input          : In std_logic;
  q_out            : Out std_logic;
  q_not_out        : Out std_logic);
End Component DffSynEa;

Begin

DffSynEa1: DffSynEa
Port Map (
  reset            => reset,
  clk              => clk,
  enable           => enable,
  d_input          => d_input,
  q_out            => q_out,
  q_not_out        => q_not_out);

create_twenty_Mhz: Process
Begin
Wait For twenty_five_nsec;
  clk<= Not clk;
End Process;

-- assigning values to input signals

reset            <=    '0' After 35.00 nsec;

enable           <=    '1' After 100.00 nsec;

d_input          <=    '0' After 40.00 nsec,
                       '1' After 70.00 nsec,
                       '1' After 145.00 nsec,
                       '0' After 245.00 nsec;

End tb_DffSynEa;
```

A–5. Latch Design Testbench

Use this testbench to verify the latch design in Chapter 2. An example of the simulated output is shown in Figure A–5.

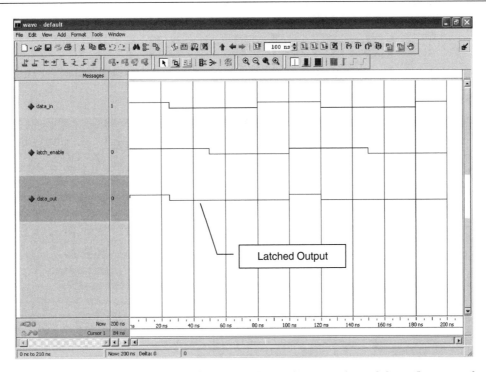

Figure A–5: Latch Simulation Outputs (Material based on or adapted from figures and text owned by Xilinx, Inc., courtesy of Xilinx, Inc. Copyright © Xilinx 1995–2008 used in Xilinx ISE WebPack™ software version 10.1.)

Run the simulation for 200.00 nsec.

```
Library IEEE;
Use IEEE.std_logic_1164.All;
Use IEEE.std_logic_unsigned.All;

Entity testbench Is End testbench;

Architecture tb_Latch Of testbench Is

Signal data_in      : std_logic := '1';    -- setting initial value
Signal latch_enable : std_logic := '1';    -- setting initial value
Signal data_out     : std_logic;
```

```
Component Latch Port (
  data_in       : In std_logic;
  latch_enable  : In std_logic;
  data_out      : Out std_logic);
End Component Latch;

Begin

Latch1: Latch
Port Map (
  data_in       => data_in,
  latch_enable  => latch_enable,
  data_out      => data_out);

-- assigning values to input signals
latch_enable    <= '0' After 50.00 nsec,
                   '1' After 100.00 nsec,
                   '0' After 150.00 nsec;

data_in         <= '0' After 25.00 nsec,
                   '1' After 80.00 nsec,
                   '0' After 120.00 nsec,
                   '1' After 180.00 nsec;

End tb_Latch;
```

A–6. Manual Shift Register Testbench

Use this testbench to verify the manual shift register design in Chapter 2. An example of the simulated output is shown in Figure A–6. This testbench can also be used for the simplified shift register design.

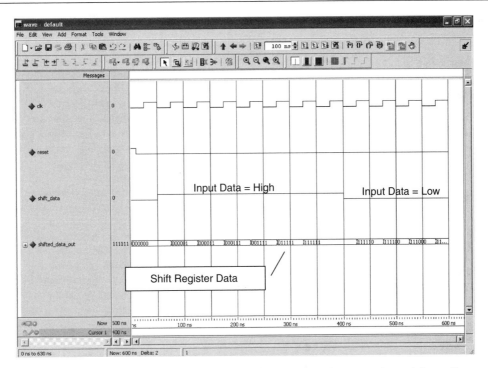

Figure A–6: Shift Register Simulation Outputs (Material based on or adapted from figures and text owned by Xilinx, Inc., courtesy of Xilinx, Inc. Copyright © Xilinx 1995–2008 used in Xilinx ISE WebPack™ software version 10.1.)

Run the simulation for 600.00 nsec.

```vhdl
Library IEEE;
Use IEEE.std_logic_1164.All;
Use IEEE.std_logic_arith.All;
Use IEEE.std_logic_unsigned.All;

Entity testbench Is End testbench;

Architecture tb_ShiftRegister Of testbench Is
Signal clk               : std_logic := '0' ; -- setting initial value
Signal reset             : std_logic := '1' ; -- setting initial value
Signal shift_data        : std_logic := '0' ; -- setting initial value
Signal shifted_data_out: std_logic_vector ( 5 Downto 0);

Component ShiftRegister
Port (
  clk                  : In std_logic;
  reset                : In std_logic;
  shift_data           : In std_logic;
  shifted_data_out     : Out std_logic_vector (5 Downto 0));
End Component ShiftRegister;
```

```
Constant twenty_five_nsec: time := 25 nsec;

Begin

U1: ShiftRegister
Port Map (
  clk                    => clk,
  reset                  => reset,
  shift_data             => shift_data,
  shifted_data_out       => shifted_data_out);

create_twenty_Mhz: Process
Begin
Wait For twenty_five_nsec;
  clk            <= Not clk;
End Process;

-- assigning values to input signals
reset          <=  '0' After 100.00 nsec;

shift_data     <=  '1' After 50.00 nsec,
                   '0' After 400.00 nsec;
End tb_ShiftRegister;
```

A–7. Comparator Testbench

Use this testbench to verify the comparator design in Chapter 2. An example of the simulated output is shown in Figure A–7.

Run the simulation for 100.00 nsec.

```
Library IEEE;
Use IEEE.std_logic_1164.All;
Use IEEE.std_logic_arith.All;
Use IEEE.std_logic_unsigned.All;

Entity testbench Is End testbench;

Architecture tb_Comparison Of testbench Is

Component Comparison Port (
  number_1             : In std_logic_vector (2 Downto 0);
  number_2             : In std_logic_vector (2 Downto 0);
  num1_smaller_num2    : Out std_logic);
End Component Comparison;
```

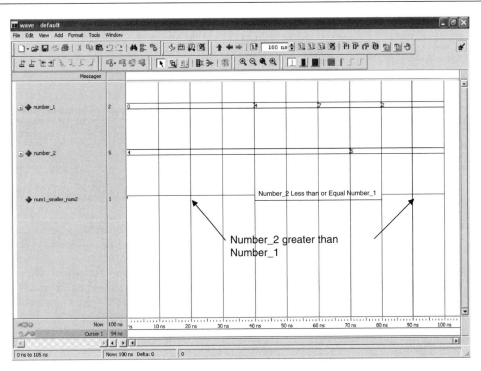

Figure A-7: Comparator Simulation Outputs (Material based on or adapted from figures and text owned by Xilinx, Inc., courtesy of Xilinx, Inc. Copyright © Xilinx 1995–2008 used in Xilinx ISE WebPack™ software version 10.1.)

```
Signal number_1          : std_logic_vector(2 Downto 0) := "000";
-- setting initial value
Signal number_2          : std_logic_vector(2 Downto 0) := "100";
-- setting initial value
Signal num1_smaller_num2   : std_logic := '0';

Begin
u2: Comparison
Port Map (
  number_1             => number_1,
  number_2             => number_2,
  num1_smaller_num2  => num1_smaller_num2);

-- assigning values to input signals
number_1      <=   "100" After 40.00 nsec,
                   "111" After 60.00 nsec,
                   "010" After 80.00 nsec;

number_2      <=   "101" After 70.00 nsec;

End tb_Comparison;
```

A–8. Binary Counter Testbench

Use this testbench to verify the binary counter design in Chapter 2. An example of the simulated outputs is shown in Figure A–8.

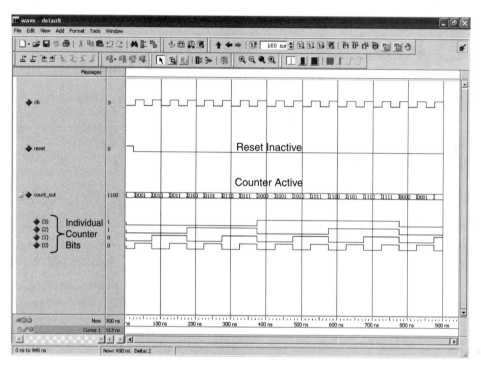

Figure A–8: Binary Counter Simulation Outputs (Material based on or adapted from figures and text owned by Xilinx, Inc., courtesy of Xilinx, Inc. Copyright © Xilinx 1995–2008 used in Xilinx ISE WebPack™ software version 10.1.)

Run the simulation for 900.00 nsec.

```
Library IEEE;
Use IEEE.std_logic_1164.All;
Use IEEE.std_logic_unsigned.All;

Entity testbench IS End testbench;

Architecture tb_BinaryCounter Of testbench Is

Signal clk          : std_logic := '0' ;
Signal reset        : std_logic := '1' ;
Signal count_out    : std_logic_vector (3 Downto 0) := "0000";
    -- setting initial state value
```

```
Constant twenty_five_nsec: time := 25 nsec;

Component BinaryCounter Port (
  clk          : In std_logic;
  reset        : In std_logic;
  count_out    : Out std_logic_vector (3 Downto 0));
End Component BinaryCounter;

Begin

U1: BinaryCounter
Port Map (
  clk           => clk,
  reset         => reset,
  count_out     => count_out);

create_twenty_Mhz: Process
Begin
Wait For twenty_five_nsec;
  clk             <= Not clk;
End Process;

-- assigning value to input signal
reset             <= '0' After 250.00 nsec;

End tb_BinaryCounter;
```

A–9. Binary Counter with Synchronous Enable Testbench

Use this testbench to verify the binary counter with synchronous enable design in Chapter 2. An example of the simulated outputs is shown in Figure A–9.

Run the simulation for 900.00 nsec.

```
Library IEEE;
Use IEEE.std_logic_1164.All;
Use IEEE.std_logic_arith.All;
Use IEEE.std_logic_unsigned.All;

Entity testbench IS End testbench;

Architecture tb_SyncBinaryCounter Of testbench Is
```

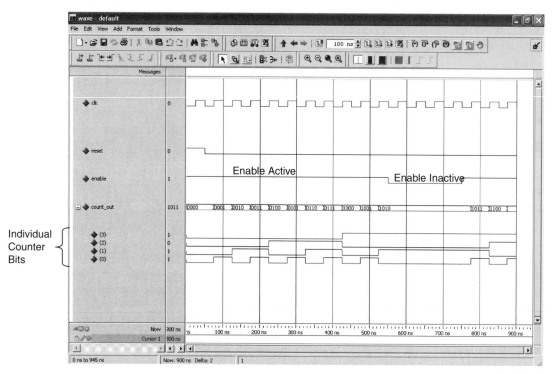

Figure A–9: Binary Counter with Synchronous Enable Simulation Outputs (Material based on or adapted from figures and text owned by Xilinx, Inc., courtesy of Xilinx, Inc. Copyright © Xilinx 1995–2008 used in Xilinx ISE WebPack™ software version 10.1.)

```
Signal clk          : std_logic := '0';
Signal reset        : std_logic := '1';
Signal enable       : std_logic := '1';
Signal count_out    : std_logic_vector (3 Downto 0):= "0000";
      -- setting initial value

Constant twenty_five_nsec: time := 25 nsec;

Component SyncBinaryCounter Is Port (
  clk          : In std_logic;
  reset        : In std_logic;
  enable       : In std_logic;
  count_out    : Out std_logic_vector (3 Downto 0)); -- output
value from counter
End Component SyncBinaryCounter;
```

Begin

U1: SyncBinaryCounter
Port Map (
  clk   => clk,
  reset   => reset,
  enable   => enable,
  count_out   => count_out);

create_twenty_Mhz: **Process**
Begin
Wait For twenty_five_nsec;
  clk   <= **Not** clk;
End Process;

--assigning values to input signals
reset   <=   '0' **After** 50.00 nsec;

enable   <=   '0' **After** 550.00 nsec,
                        '1' **After** 750.00 nsec;
End tb_SyncBinaryCounter;

A–10. Conversion Testbench

Use this testbench to verify the conversion design in Chapter 2. An example of the simulated outputs is shown in Figure A–10.

Run the simulation for 40.00 nsec.

Library IEEE;
Use IEEE.std_logic_1164.**All**;
Use IEEE.std_logic_arith.**All**;
Use IEEE.std_logic_unsigned.**All**;

Entity testbench **Is End** testbench;

Architecture tb_Convert2Integer **Of** testbench **Is**

Signal number_1   : std_logic_vector(2 **Downto** 0) :=
"000";   *-- setting initial value*
Signal number_2   : std_logic_vector(2 **Downto** 0) :=
"010";   *-- setting initial value*
Signal quotient   : std_logic_vector(2 **Downto** 0);

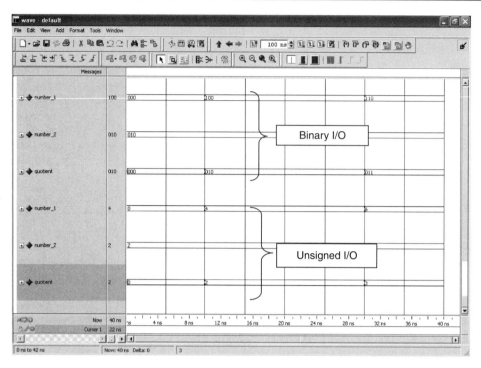

Figure A–10: Conversion Simulation Outputs (Material based on or adapted from figures and text owned by Xilinx, Inc., courtesy of Xilinx, Inc. Copyright © Xilinx 1995–2008 used in Xilinx ISE WebPack™ software version 10.1.)

```vhdl
Component Convert2Integer Port (
  number_1  : In std_logic_vector (2 Downto 0);
  number_2  : In std_logic_vector (2 Downto 0);
  quotient  : Out std_logic_vector (2 Downto 0));
End Component;

Begin

u3: Convert2Integer Port Map(
  number_1      => number_1,
  number_2      => number_2,
  quotient      => quotient);

-- assigning values to input signal
number_1        <=   "100" After 10.00 nsec,
                     "100" After 20.00 nsec,
                     "110" After 30.00 nsec;
End tb_Convert2Integer;
```

Index

Page numbers followed by "*f*" indicate figures, "*t*" indicate tables, and "*b*" indicate boxes and formulas.